HOME HEATING AND AIR CONDITIONING SYSTEMS

James L. Kittle
Line Drawings by
Gary McKinney

To the homeowner/handyperson:
To assist you in doing a better, safer job
of installing, repairing, and maintaining
your equipment.

FIRST EDITION
THIRD PRINTING

Library of Congress Cataloging-in-Publication Data

Kittle, James L., 1913-
Home heating and air conditioning systems / by James L. Kittle.
p. cm.
ISBN 0-8306-9257-6 ISBN 0-8306-3257-3 (pbk.)
1. Dwellings—Heating and ventilation—Amateurs' manuals.
2. Dwellings—Air conditioning—Amateurs' manuals. I. Title.
TH7224.K58 1990
697—dc20 89-29159
CIP

TAB Books offers software for sale. For information and a catalog, please contact TAB Software Department, Blue Ridge Summit, PA 17294-0850.

Acquisitions Editor: Kimberly Tabor
Production: Katherine G. Brown

Contents

Acknowledgments

The following companies have provided illustrations and text for this book. Their cooperation and courtesy are greatly appreciated.

American Gas Association, Arlington, Virginia.
The Beckett Corporation, Elyria, Ohio.
Blue Ray Systems, Upper Darby, Pennsylvania.
Bryan Steam Corporation, Peru, Indiana.
Burnham Corporation, Lancaster, Pennsylvania.
Carrier Corporation, Syracuse, New York.
Consolidated Dutchwest, Inc., Plymouth, Massachusetts.
Electric Furnace-Man Div., Emmaus, Pennsylvania.
Energy Vent, Inc., Dayton, Ohio.
Honeywell Inc., Minnetonka, Minnesota.
Hydrotherm, Northvale, New Jersey.
ITT Bell & Gossett, Morton Grove, Illinois.
Intertherm Nordyne, A Nortek Co., St. Louis, Missouri.
Klein Tools, Chicago, Illinois.
Lopi® International, LTD., Kirkland, Washington.
Monarch Manufacturing Works, Inc., Philadelphia, Pennsylvania.
Mueller Brass Company, Port Huron, Michigan.
Research Products Corporation, Madison, Wisconsin.
Ridge Tool Company, Elyria, Ohio.
Suburban Manufacturing Company, Dayton, Tennessee.
Suntec Industries, Inc., Rockford, Illinois.
Toastmaster, Inc., Columbia, Missouri.
The Trane Company, Tyler, Texas.
Vaco Products Div., Chicago, Illinois.
Weil McLain, A Marley Company, Michigan City, Indiana.

Westwood Products, Inc., Perth Amboy, New Jersey.
White-Rodgers Div. Emerson Electric Company, St. Louis, Missouri.
Woodstock Soapstone Company, Inc., West Lebanon, New Hampshire.
Yukon Energy Corporation, St. Paul, Minnesota.
Bryant Air Conditioning, Indianapolis, Indiana.
McDonnell & Miller, ITT, Chicago, Illinois.
Infloor Heating Systems, Hamel, Minnesota

Introduction

MANY HOW-TO BOOKS ON HEATING AND AIR CONDITIONING SYSTEMS ARE ON the market today. Although many are well written and illustrated, others have poor or stylized drawings that do not resemble actual objects. More important, most are not written by people who have "worked with the tools," but by writers who obtain information from journeymen, inspectors, engineers, and others who are knowledgeable in their field. Often, these types of how-to books are based on second hand accounts, and, at best, are not always correctly written or explained.

This book is written for the homeowner/handyperson who is interested in maintaining, altering, or installing heating or cooling equipment in their own home. Every government entity, whether township, city, or county, has rules that prohibit anyone who is not licensed from working on electrical, plumbing, heating, and cooling equipment other than in their own home.

I have written this book by drawing on my many years of practical experience, not as a do-it-yourselfer, but as a journeyman with hands-on experience in all of the topics covered in this book. Every procedure, operation, repair, new installation, rebuilding, and removal and replacement I have, at one time or another, done, and not under ideal laboratory conditions. *Home Heating and Air Conditioning Systems* contains complete, detailed instructions and illustrations so that you can get the work done correctly, in accordance with local regulations.

In this book, I'll show you how to do a procedure the easiest, best, and fastest way. Over the years, individual journeymen have developed methods and special procedures that are actually shortcuts, improvements, and safer ways to perform necessary work. Some of these procedures should *not* be done by a handyperson or homeowner, because they might be hazardous to equipment, such as a heat pump or air conditioning compressor. I recommend that these complicated procedures be done only by a trained technician in that particular field. Many of the procedures are illustrated, and diagrams are furnished to accompany the written instructions. Some of the procedures that are explained in this book are detailed, and although I've tried to write everything in clear, easy-to-understand terms, it might be necessary for you to read the instructions more than once. Follow directions sequentially, and do not skip steps. Doing procedures out of order could result in damaged equipment.

Chapters 1 and 2 cover safety precautions and selecting and using tools correctly. Be sure to read these chapters before you begin any work so that you are aware of the potential hazards of handling flammable materials or explosive gases such as propane and natural gas.

Chapters 3 and 4 explain how to detect faulty installations and how to evaluate gas- and oil-fired furnaces. Chapter 5 explains how different heating systems work and provides practical advice for choosing a new system or upgrading an older one. In Chapters 6 and 7, you'll learn how to install an oil-fired furnace and replace a boiler.

There is also information on how to install central air conditioning and window units, as well as a complete chapter on using a heat pump for supplemental heat or alone. There is also a complete chapter on servicing heating and air conditioning systems as well as a time-saving troubleshooting index.

For the most part, all of the work is within the ability of the average handyperson, do-it-yourselfer. Exceptions might be making the final connection to a live electrical panel, or replacing a burned-out air conditioning compressor. Other than these few instances, all of the work explained and illustrated in this book can easily be done by the do-it-yourselfer.

I hope you will find this book informative and helpful in enabling you to keep your mechanical heating and cooling equipment maintained in excellent condition.

1

Safety

SAFETY IS AN IMPORTANT CONSIDERATION IN ALL WALKS OF LIFE, AND ESPEcially in the home where there are young children. Keep safety in mind at all times when working on any mechanical equipment such as heating and air conditioning appliances. Do not work on equipment when it is in operation, even though the equipment has shut itself off. In this mode, the equipment can start at *any* time and cause injury to anyone attempting to service or adjust it.

FIRE HAZARDS

Fire that is controlled, as in heating equipment, *is* safe. The nature of fire must be thoroughly understood, however, even when working with heating equipment. Most materials will burn when their temperature is raised to the ignition temperature of that material. Paper and wood shavings ignite at a very low temperature. Wood has a much higher ignition point.

Number 2 fuel oil is not easily ignited in liquid form, but when sprayed into the combustion chamber of an oil-fired furnace or boiler, ignites easily. This is controlled combustion, and is safe. Natural gas and liquified petroleum gas (LPG), also called "bottled gas," are extremely flammable and must *always* be treated with respect. Both of these gases are under pressure and the piping must be gastight to prevent leaks. Natural gas is lighter than air and tends to rise. LPG is heavier than air and, if leaking, will flow along the floor or into a crawl space under a building for great distances. Odor-producing substances are added to both gases for leak detection purposes. A leak of either gas presents a *very hazardous condition*! An accumulation of either gas, if ignited by a spark or flame, can demolish a large building.

The only safe way to find a gas leak is with soap bubbles or leak detection solution (similar to a child's bubble liquid). Paint *all* joints with the thick ropey solution. Bubbles will form at the location of a leak. When testing a new or repaired gas line, shut off all gas-using equipment, and pilot lights. As an additional check on leaking gas, observe the one-half or one-cubic foot dial on the meter. This dial is separate from the usage dials and is usually below or to one side of these dials. If the hand moves, there is a leak. If the hand does not move, the system is gastight.

GAS UTILITIES

All public utilities are regulated by a state Public Service Commission. Gas is supplied to the utility lines at high pressure, which is reduced to about seven inches water column (WC), about a half pound, by a pressure regulator (reducer) near the meter. Another regulator on the furnace, ahead of the automatic gas valve, reduces the pressure further to three inches WC. If this is a new installation or an extensive remodeling project, you will have to contact the gas utility before you start working. You will also need a permit from the governing authority, city, township, or other administrative body.

Gas connections, such as gas-fired clothes dryers and gas-fired barbecues are sometimes piped using 3/8-inch O.D. copper or aluminum tubing. NOTE: *All* tubing connections on gas lines *must* be the flare type. This type of connection is the most leak-proof and is very reliable.

OIL-FIRED FURNACES

Oil-fired equipment is much different than gas-fired types. Oil leaks are also not as dangerous as gas leaks. An oil leak will show as an oily spot on the floor, or as a wet line near a fitting (a crushed line may also leak). Leaking fuel oil *is* a fire hazard and must be repaired as soon as possible. Oil lines of copper (do not use aluminum) are connected using flare fittings, as are gas lines. CAUTION: *Always* run oil, gas, and water lines so as to minimize the chance of damage to them. It is better to cut a trench in the concrete basement floor, bury the oil line, then cement over the line, rather than run the line where it would be subject to damage.

Formerly, oil-burning furnaces were required to have a red shutoff switch located near the entrance to the basement so that the burner could be shut off in case of emergency. This is no longer required. A fire extinguisher is now recommended for this location. The fuel oil tank, if located in the basement, must be at least 10 feet from the oil burner. Most building codes require this minimum distance.

BUILDING CODES

Before you can install new heating equipment, you must obtain a permit from the building department of the governing body for that area. Talk with the inspector and bring an outline of your installation, showing the location of the burner and the oil tank. Outside tanks are satisfactory, except that water in the oil tank can freeze during winter and cause the burner to stop.

Many times, a substandard installation results from poor workmanship, ignorance of standard practices, and violations of applicable building codes. Electric wiring is one example. Polarity of wires are not maintained; the white and black wires are interchanged, posing the danger of shock. The National Electrical Code (NEC) makes it mandatory that the white wire (neutral) cannot be broken, except when all live (hot) wires are broken (disconnected) simultaneously with the neutral wire. Another rule that is mandatory: The green insulated or bare wire must be continuous, from the service entrance panel entirely through all of the wiring circuits. This is the bare ''ground wire'' found in all electrical cable. These requirements might seem minor, but they are vitally important in practice for the safety of people who will be coming in contact with the electrical equipment.

2

Tools

THE AVERAGE HOMEOWNER/HANDYPERSON WILL HAVE MANY OF THE STANDARD hand and power tools described in this chapter. People who specialize in woodworking projects might not own many of the tools used for working on mechanical equipment. Tools that you will only use occasionally can be borrowed or rented. Many fine tools are often on sale at reduced prices and are worth buying, *if* you have a future use for them.

HAND TOOLS

Pliers, screwdrivers, and small wrenches are usually on hand. Because quality tools ''work right'' and last, literally forever, they are the best buy, even though the cost is more. Brand name tools, made by reputable manufacturers will give excellent service and will usually be replaced at no charge if broken. See Figs. 2-1 through 2-7. The pliers you will need are ''waterpump pliers,'' and Channellock is the best. Electrician's pliers, side cutting lineman's and diagonal type can be either Klein or Channellock. They also make good long-nose and other types of pliers. You will need several screwdrivers: a small pocket size, a medium size, and a medium large size. A Phillips medium size screwdriver will fit most Phillips screws found on heating and cooling equipment. Small sets of open-end wrenches are nice to have, but adjustable wrenches are my favorites. The six-inch, eight-inch, and 10-inch sizes will fit almost every nut and bolt found on heating and air conditioning equipment.

Standard hacksaws have an adjustable frame that will accept 10- or 12-inch blades. The 12-inch blade allows a longer stroke, which can be less tiring when doing a great deal of metal sawing. The ''keyhole'' type allows sawing in confined spaces not large enough to accept the large saw frame. This saw has a tapered blade fastened at only one end in a pistol-grip handle. A standard hacksaw blade will substitute for the keyhole saw. Tape one end for use as a handle. The teeth must point forward when using the blade alone and in the frame. Standard blades come in various "teeth per inch," such as 16, 24, and 32 teeth per inch. The thickness of the metal will determine the number of teeth. Thin brass tubing on a sink drain line will catch the teeth of a 16 tooth blade. To better cut sheet metal, use sheet metal snips. The newer style of snips, known as "aircraft snips," are usually better to use than the large, older type, except for long, straight cuts. Wiss makes the best quality. Aircraft snips are available in "cuts right," "cuts left," and "cuts straight." These types of snips are favored by sheet metal workers for cutting holes in ductwork.

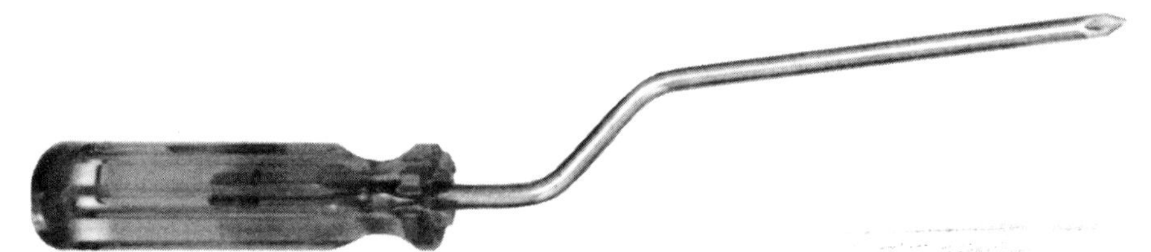

Fig. 2-1. *A Rapidriv screwdriver.* Courtesy of Vaco Products Division.

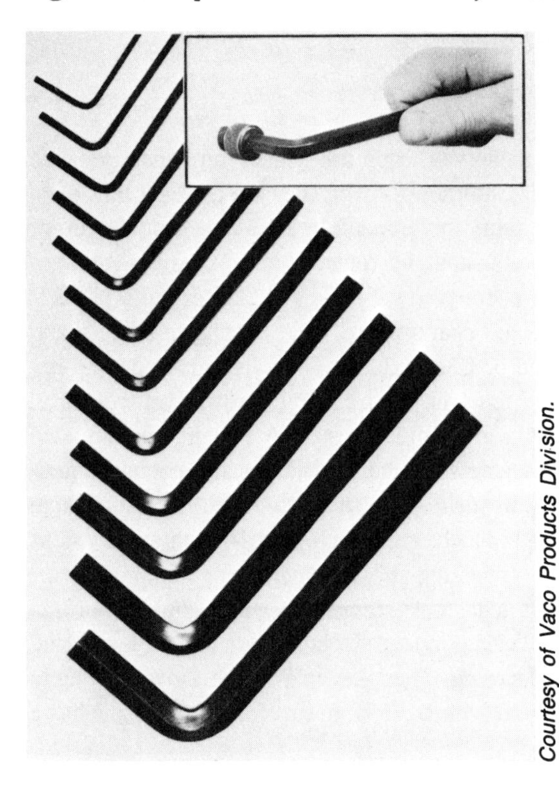

Courtesy of Vaco Products Division.

Fig. 2-2. *Hex keys come in long and short series. Pictured here are short series hex keys.*

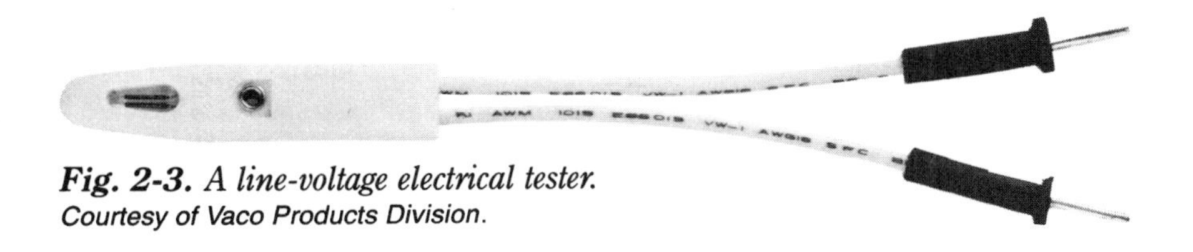

Fig. 2-3. A line-voltage electrical tester.
Courtesy of Vaco Products Division.

Nut drivers are sometimes more convenient to use than an adjustable wrench. Kits of assorted sizes of sockets, with a separate handle or separate sockets, each with its own handle, are available.

Bolt thread taps and dies are sometimes necessary. A set can be started for a small cost. Buy the size tap or die you need now and the appropriate handle to match. If you continue to buy items as you need them, you will soon have a starter set. Many trade people do this because not all sizes are used by all trades.

Locking pliers are very handy for use as a small vise or for loosening bolts and screws having damaged heads. Vise-Grip by Petersen Company is the best for long life and fine operation. All top quality tools have imitations and look-alikes on the market. Buy the best for long life and satisfaction.

POWER TOOLS

Most home workshops have an electric drill. If you don't already own one, the best all around size is the 3/8-inch capacity, variable speed, reversing electric drill. This type is very popular on the job, and in the home workshop. The best top brands are: Black and Decker, Skil, Porter-Cable, and Craftsman (Sears). All are made in the U.S.A. An electric saber-saw is convenient for sawing openings in ductwork (use a fine-tooth blade). Drill bits for the electric drill should be designated "high speed steel," or "HS" might be stamped on loose drill bits. Do not buy "carbon" drill bits, these are for wood and plastic only. Carbide drill bits are for drilling into masonry, such as Carboloy. Hammer-type electric drills are available to expedite drilling, but for household use, the standard electric drill is satisfactory.

You will need several heavy-duty extension cords for working on furnaces and air conditioners. Some come with a trouble light, which has one or two outlets in the handle for plugging in an electric drill or other tool. A 50-foot cord should have 16-3 wire and good insulation for heavy-duty use. A 100-foot cord should have 14-3 wire to overcome any voltage drop at the far end of the cord. These cords are "grounded," thus the plug has three prongs and the connector has three holes. This for use with grounded power tools having a three-prong plug. Do not use the "three-to-two" adapter, as a true ground is not assured.

In the last few years, "double insulated" tools have become popular. These tools have only a 2-wire cord and do not require a 3-wire cord. They are listed by the Underwriters Laboratories (UL). Caution: When these power tools need repairs, *always* take them to the manufacturer's service center. Proper repair can only be made through these centers.

Figure 2-4 shows a terminal crimper for crimping terminals on wire ends. Figure 2-5 shows representative terminal shapes for various applications. Use a nylon tie, such as the one shown in Fig. 2-6 for tying bundles of wires, tubing, or other items.

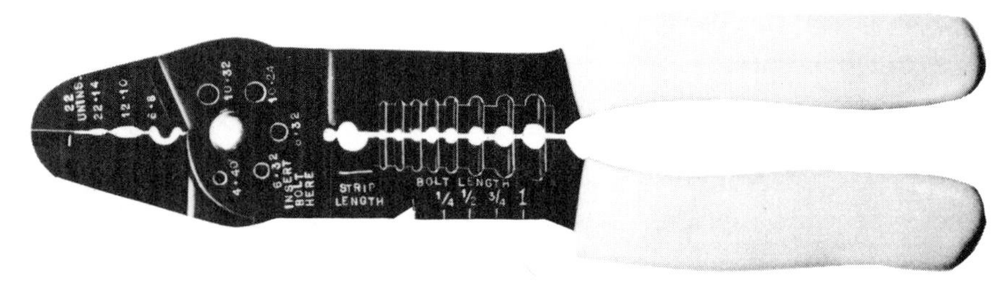

Fig. 2-4. *An all-purpose crimping, bolt cutting, and insulation stripping tool, complete with cushioned handles.* *Courtesy of Vaco Products Division.*

PIPEFITTING TOOLS

Tools for working with threaded pipe include pipe wrenches, a pipe cutter, a pipe reamer, pipe dies, and a vise. You might have several small pipe wrenches, such as 10 inch and 14 inch, and a bench vise that has pipe jaws. Other tools and larger pipe wrenches for working on boilers can be easily rented. Be sure to check the condition of the tools from rental stores, they might not be in good condition. See Figs. 2-7 through 2-23.

Threading pipe is simple if you follow the directions. Cut the pipe to length, using the pipe cutter and vise (see Fig. 2-8). Ream the cut end with the reamer (see Fig. 2-7), the burr should be completely removed so that there is free flow of liquid or gas through the pipe. A pipe cutter in good condition will cut a perfect groove around the pipe, one in poor condition can tend to "cut a thread" instead of tracking properly. This is caused by a worn shaft or cutter wheel. Check this before leaving the rental store. If the cutter starts a perfect groove, continue the cut, swinging the handle over the pipe and away from you, then back and around under the pipe. Tighten the handle slightly for each turn. Support the piece being cut off so that it does not fall and hurt you. A long piece should be supported at the far end. The piece and the cutter can fall and land on your foot. Be careful!

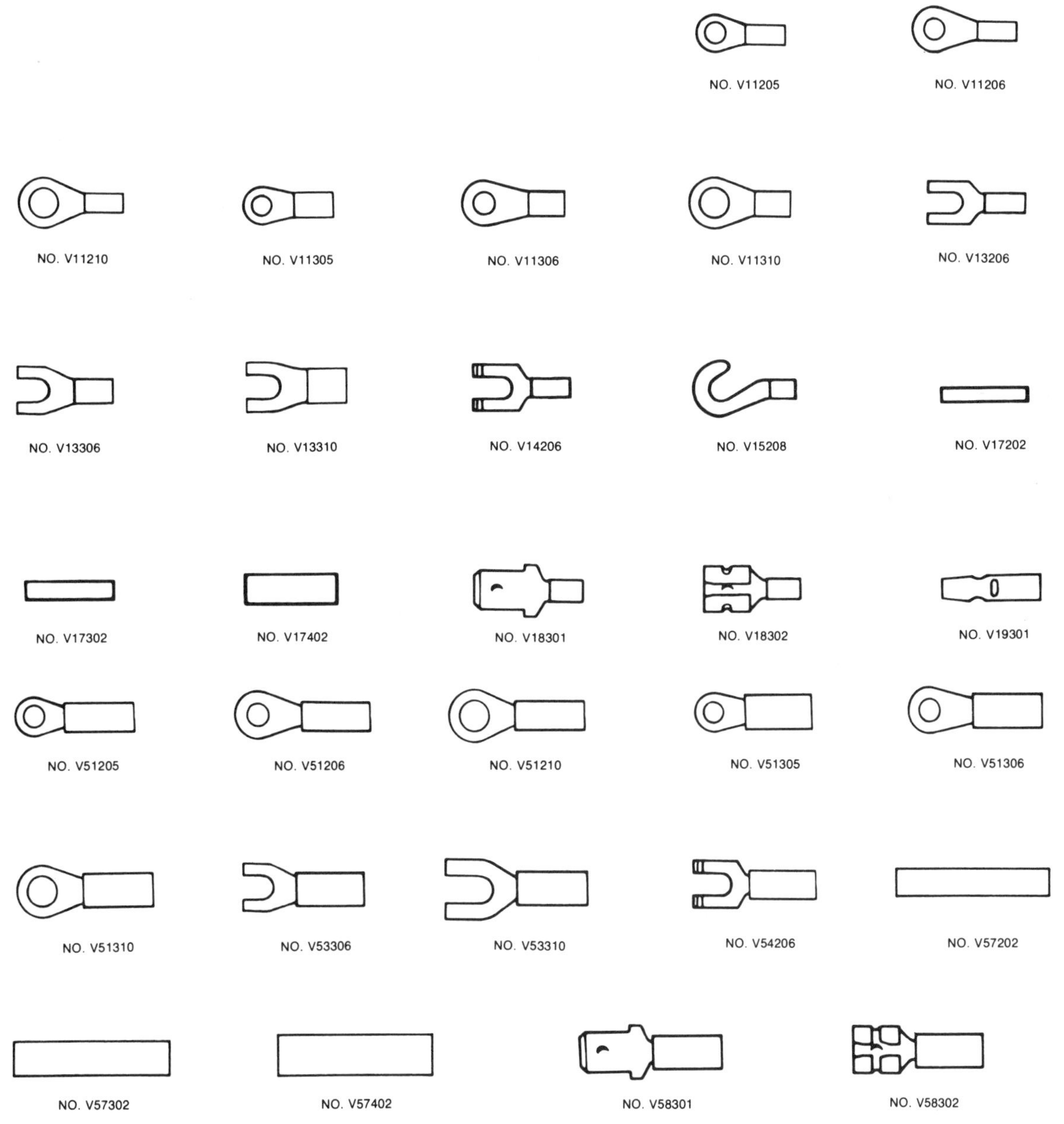

Fig. 2-5. *Solderless, crimp-type terminals.* Courtesy of Vaco Products Division.

Fig. 2-6. *A nylon tie used for bundling wires, cables, etc.* Courtesy of Vaco Products Division.

When threading pipe, select the correct die for the pipe size (see Fig. 2-8). It is easy to confuse $1^1/_4$ inch and $1^1/_2$ inch. Put the die on the cut and reamed end of the pipe, push on the die head as you attempt to start the die cutting. Some dull pipe cutters can leave a raised burr on the outside of the pipe, you might have to file this burr off so that the die will start. Use pipe cutting oil, obtainable from hardware stores, to lubricate the die as you use it. All pipe dies are "right hand" unless marked on the die as "L.H." See Fig. 2-13.

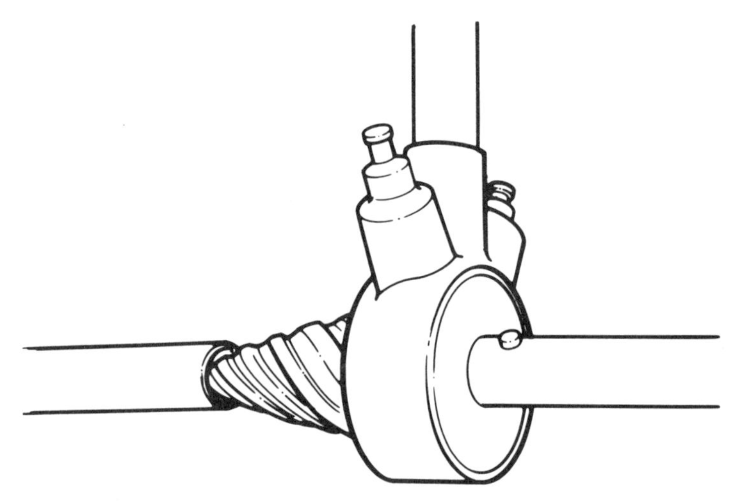

Fig. 2-7. *A spiral reamer (also made in straight style), used to ream burr from pipe after cutting.*

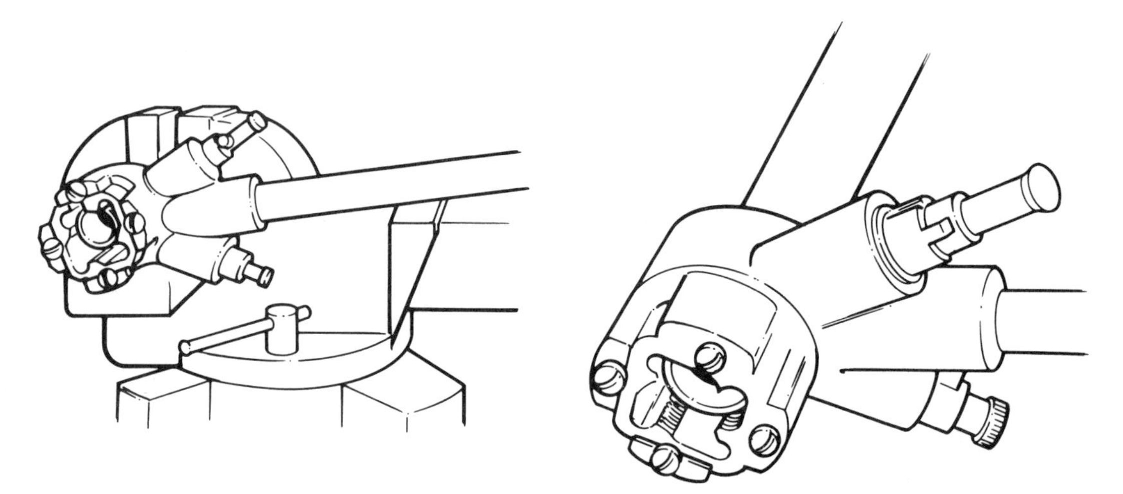

Fig. 2-8. *A pipe die cutting thread on a pipe end, nearly completed, and a pipe die started onto the pipe end. It must be first pushed to start die onto the pipe. Later, it will feed itself forward.*

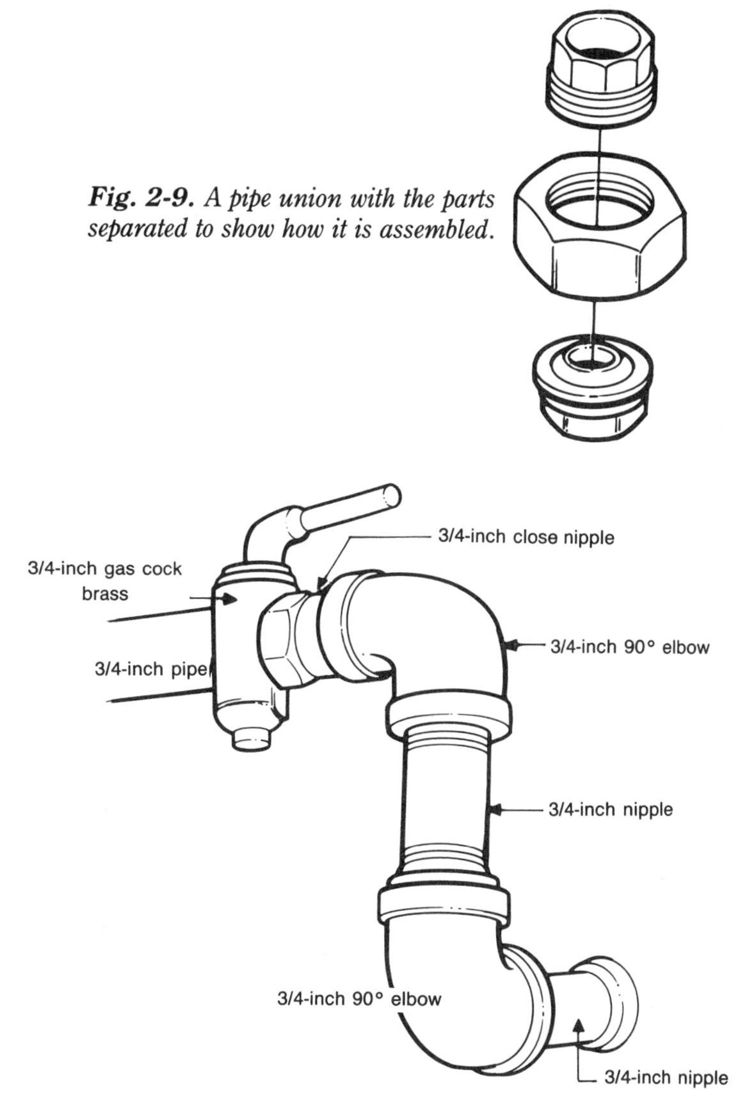

Fig. 2-9. *A pipe union with the parts separated to show how it is assembled.*

Fig. 2-10. *The piping arrangement of a gas line, including a gas cock.*

Many times one fitting can be tightened on the pipe while it is still in the vise (see Fig. 12-14). Use either pipe thread compound or Teflon tape. This tape is clean and easy to use, and is also sold at hardware stores. The only drawback is that the tape is not adhesive, so the starting end must be held while stretching it around the threads. For $^{1}/_{2}$-inch and larger pipe, the tape must be wrapped around the threads two times. Smaller pipe threads need only one turn.

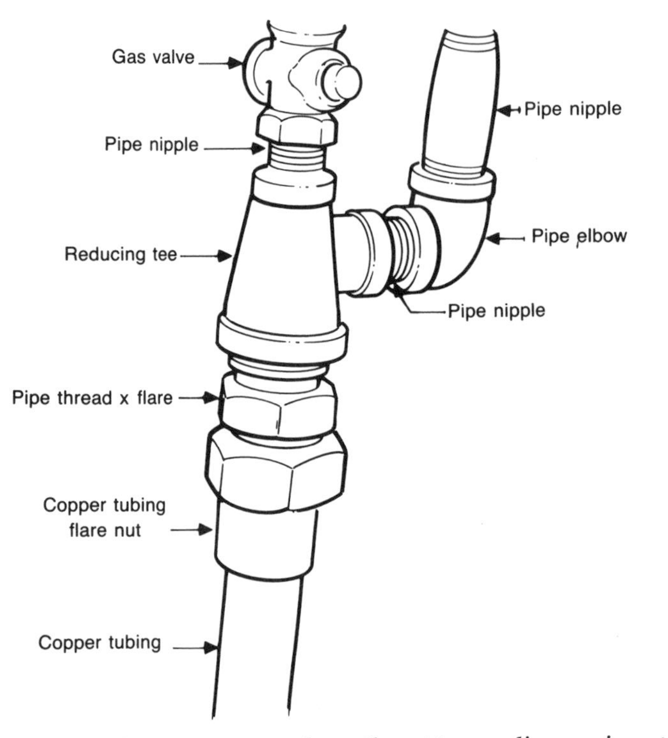

Fig. 2-11. This piping arrangement shows the copper gas line coming up through the concrete floor. The tubing is connected using a flare nut.

COPPER AND ALUMINUM TUBING TOOLS

The best method for joining copper or aluminum tubing is the flare type connection. This consists of a fitting with an external thread. At the outer end of the thread is a slanting face at a 45° angle. A flare nut has an internal thread that mates with the external thread of the fitting, and also has a mating surface of 45°. The flared tubing is clamped securely between these two surfaces, making a very tight connection. See Figs. 2-10 and 2-11 and Figs. 2-20 through 2-23.

To make a perfect flare, the tubing must be cut, using a tubing cutter (similar to a pipe cutter), and reamed with the reamer attached to the cutter (see Fig. 2-17). The end of the tubing is clamped in the proper size opening in the flare block and positioned so that the tubing end is slightly above the flare block surface, about 1/16 inch or less, then clamped tightly by tightening the wing nuts. See Fig. 2-16. A flaring tool is slid over the block and centered over the tubing. This tool consists of a cone on a threaded shank with a cross handle and a body to hook over the flare block. Turning the cross handle forces the cone into the tubing open end and forces the end into a 45° flare shape to match with the fitting and nut. NOTE: The flare nut must ALWAYS be slipped on the tubing (with the open end of the nut facing the end of the tubing) BEFORE making the flare. If the nut is not slipped on the tube first, there *may* not be enough length remaining to cut the flare off, slip the nut on and make a new flare. Some flaring sets have a captive tool on the block to prevent loss.

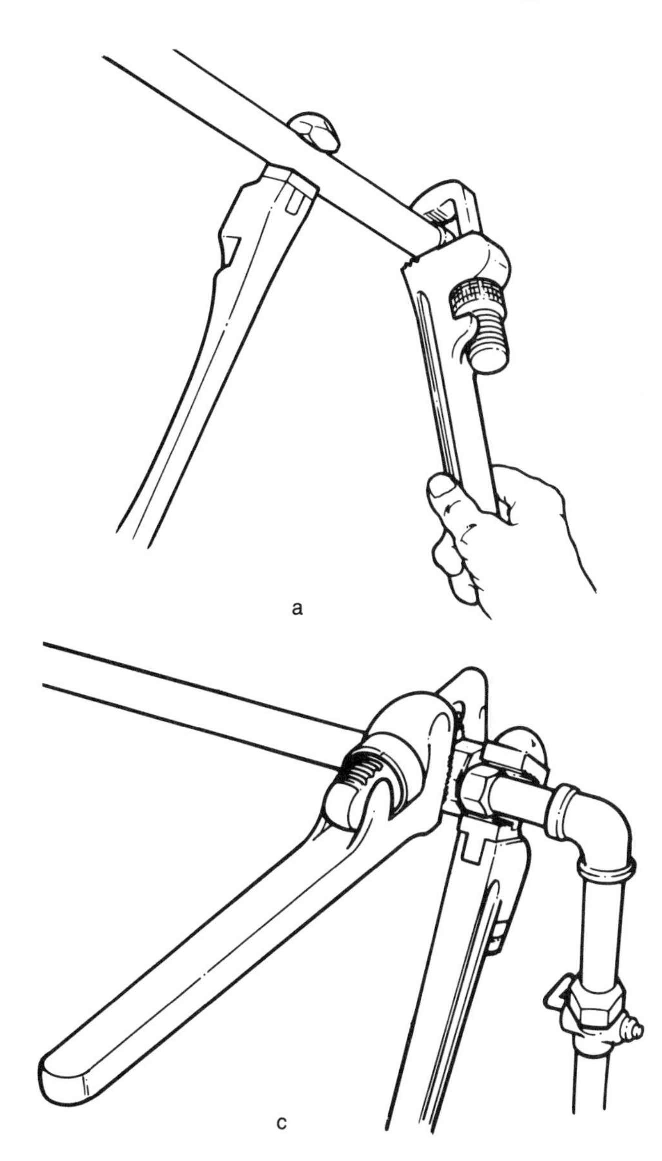

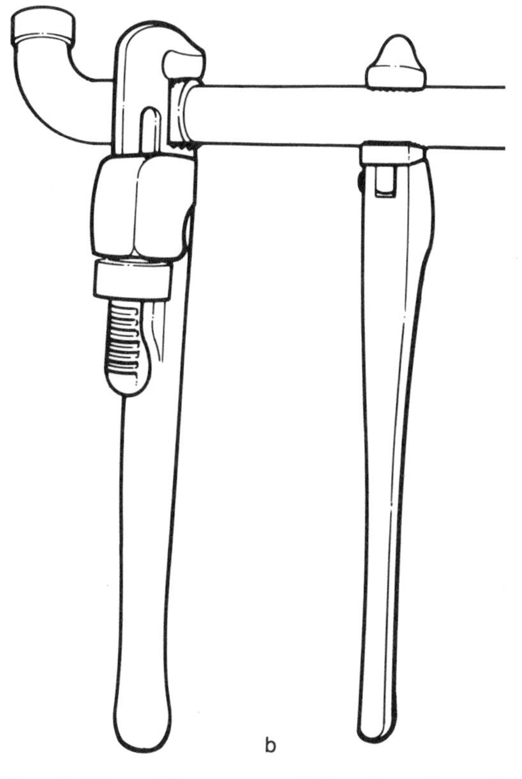

Fig. 2-12. *a. To remove a fitting from the end of a pipe, the pipe and the left wrench should be set on the floor. The right wrench is pushed down to free the fitting. b. To tighten a fitting onto the pipe, reverse the wrenches on the pipe and fitting. The pipe and right wrench rest on the floor, the left wrench is pushed down to tighten the fitting. c. Tighten the pipe union as shown, and remember to always use a backup wrench.*

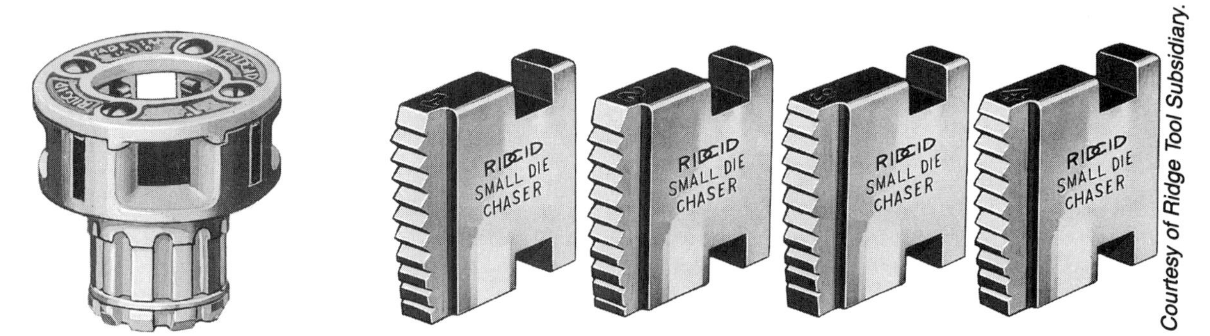

Fig. 2-13. *A pipe die holder and four dies, complete with dies installed.*

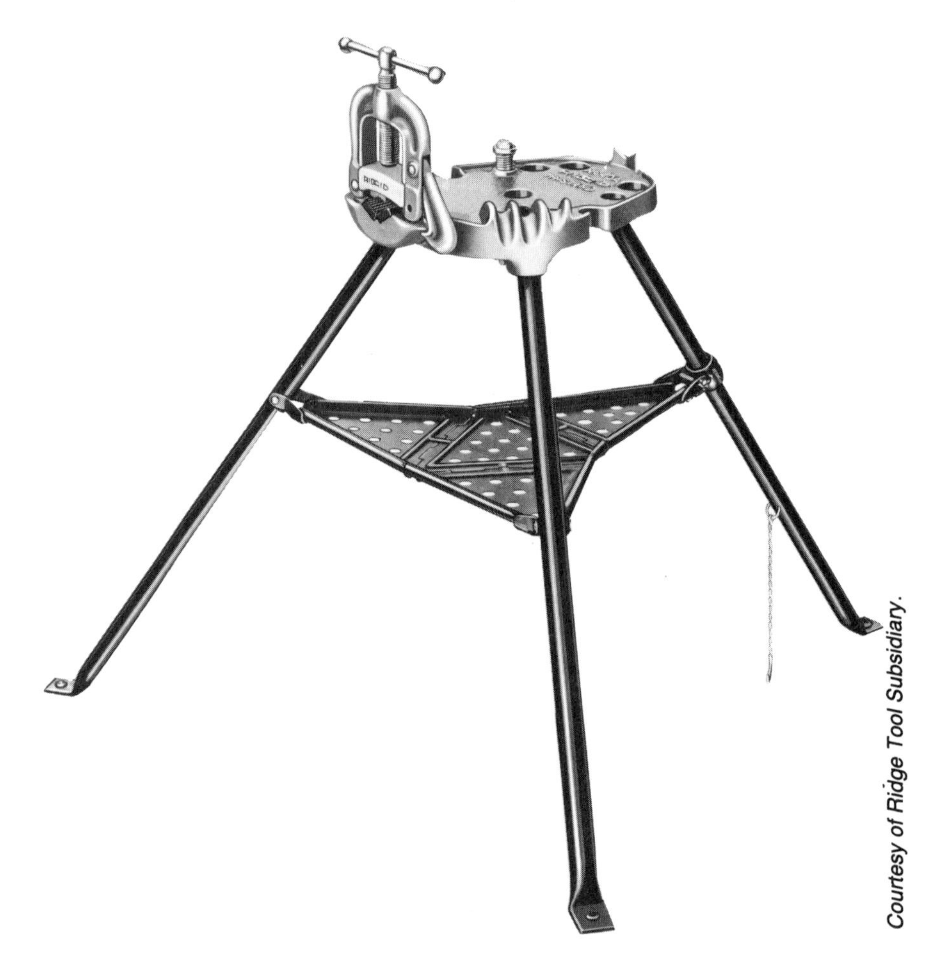

Fig. 2-14. *A pipe vise mounted on a three-legged pipe stand.*

Soldering Copper Tubing and Fittings

Figure 2-15 shows some of the tools you'll need for soldering tubing and fittings. In addition to flare connections, rigid copper pipe and copper fittings are, more often than not, used instead of flare connections. Modern building

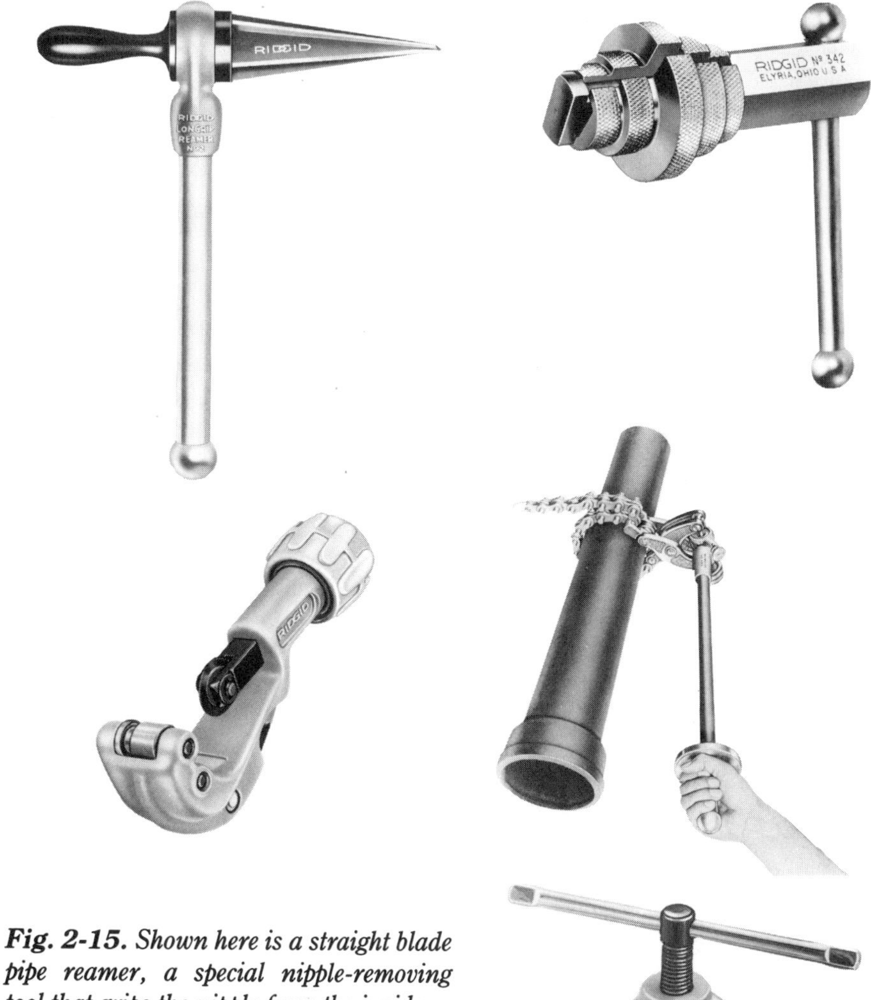

Fig. 2-15. *Shown here is a straight blade pipe reamer, a special nipple-removing tool that grips the nipple from the inside; a copper and aluminum tubing cutter, including a reamer for reaming inside of tubing after cutting; a cutter for cutting cast-iron soil pipe, including multiple cutting wheels for working in close quarters; and a flaring tool to flare various sizes of tubing.* Courtesy of Ridge Tool Subsidiary.

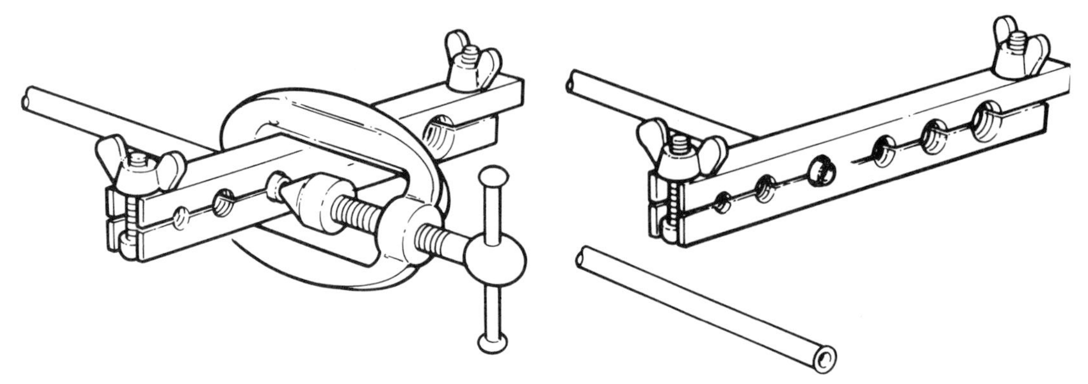

Fig. 2-16. A different type of flaring tool. The tubing is clamped in the mandrel, ready for flaring. There is also a completed flare on tubing.

construction has used copper pipe and fittings for many years with excellent success. Most domestic plumbing is rigid copper pipe and copper fittings joined with solder and flux. See Figs. 2-17 through 2-19. Solder joints are made using ''95-5'' solder (95 percent tin-5 percent antimony amalgam) for potable water lines. This is mandated by federal law (as a precaution against lead poisoning). 50 percent tin and 50 percent lead solder has been banned. This law was passed in 1986 to eliminate lead in water. Copper tubing must be prepared before soldering by cleaning with sand cloth (similar to emery cloth but made with fine sand instead). Polish the tubing end and the inside of the fitting with the sand cloth until they shine (see Fig. 2-18). Do not touch the cleaned ends with the fingers, because the solder might not take. Apply the flux in a thin, even coat using a small ''acid brush,'' a small brush with a metal handle (see Fig. 2-19).

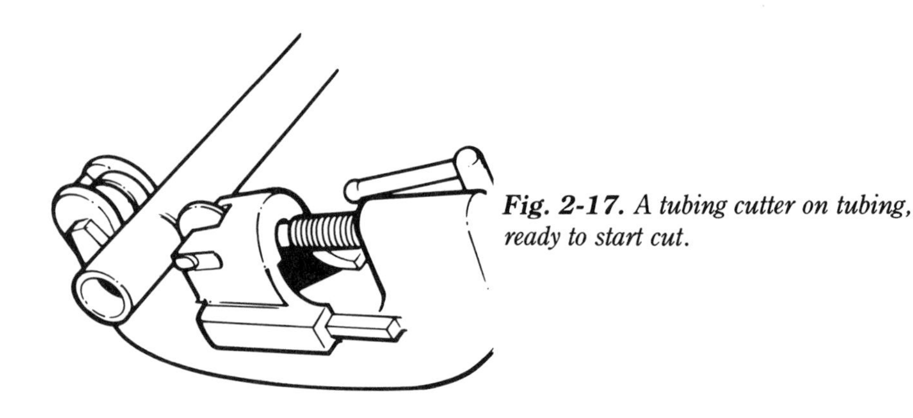

Fig. 2-17. A tubing cutter on tubing, ready to start cut.

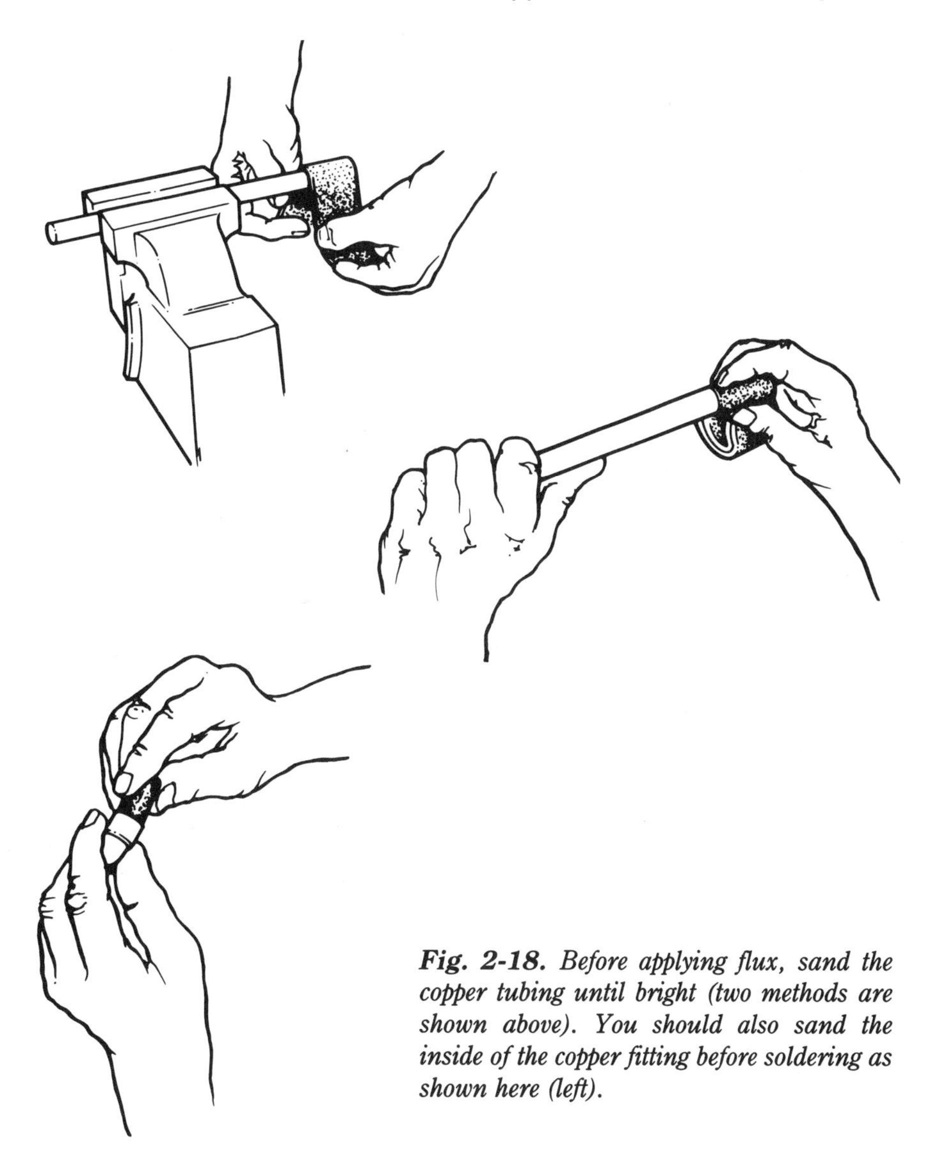

Fig. 2-18. Before applying flux, sand the copper tubing until bright (two methods are shown above). You should also sand the inside of the copper fitting before soldering as shown here (left).

Soldering flux and solder can be found in hardware stores. Be sure to ask for "95-5" solder if you are soldering domestic potable water lines. Portable propane torches are sold in hardware stores for about $10 to $15, including the propane tank and torch complete.

When the joint is assembled, heat the area of the socket and the tubing end inside the socket. Play the flame back and forth on this area, heating both the front and back equally. Do not hold the flame in one place as you can burn the flux and the copper. Straighten a portion of the solder roll, about six inches,

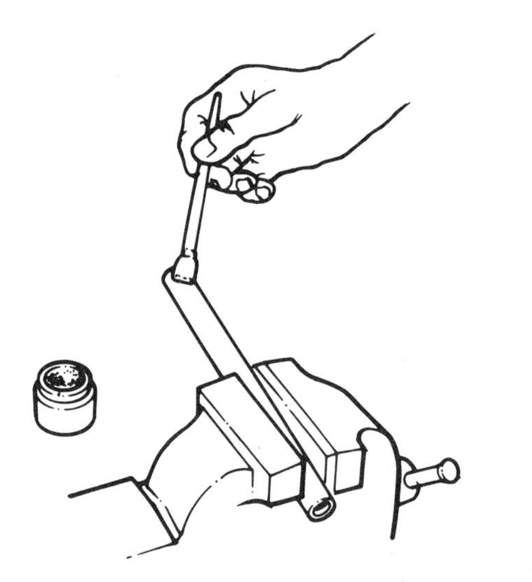

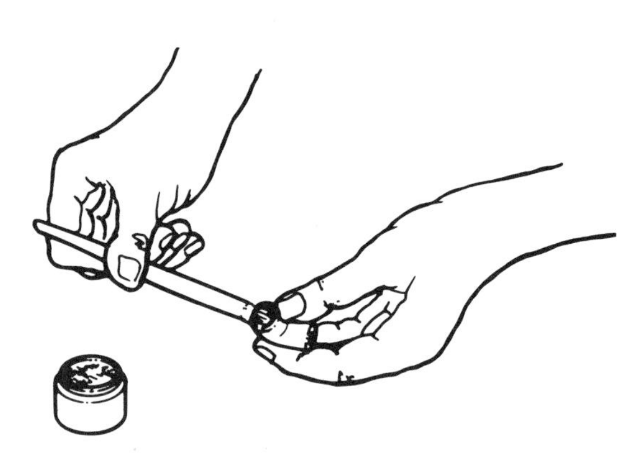

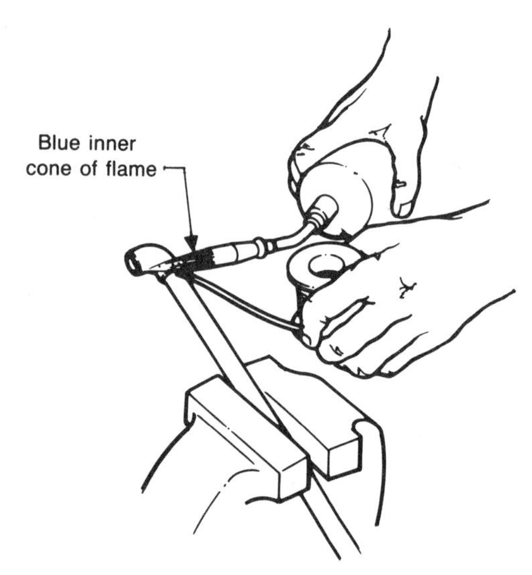

Fig. 2-19. *After sanding, apply soldering flux with an acid brush to the outside and inside of the fitting (above). Once the soldering flux is applied, solder the fitting onto a length of copper tubing as shown (left).*

and touch the end to the fitting edge. If the joint is at the right temperature, the solder will melt and flow into the space between the pipe and the fitting socket. Joints that are upside down are easily soldered, because the solder will flow in the small annular space between the two parts of the joint, regardless of the force of gravity. It is important to heat the joint evenly all around so that the molten solder will flow and completely fill the annular space. Figures 2-20 through 2-23 are a representative of copper and brass fittings for soldering.

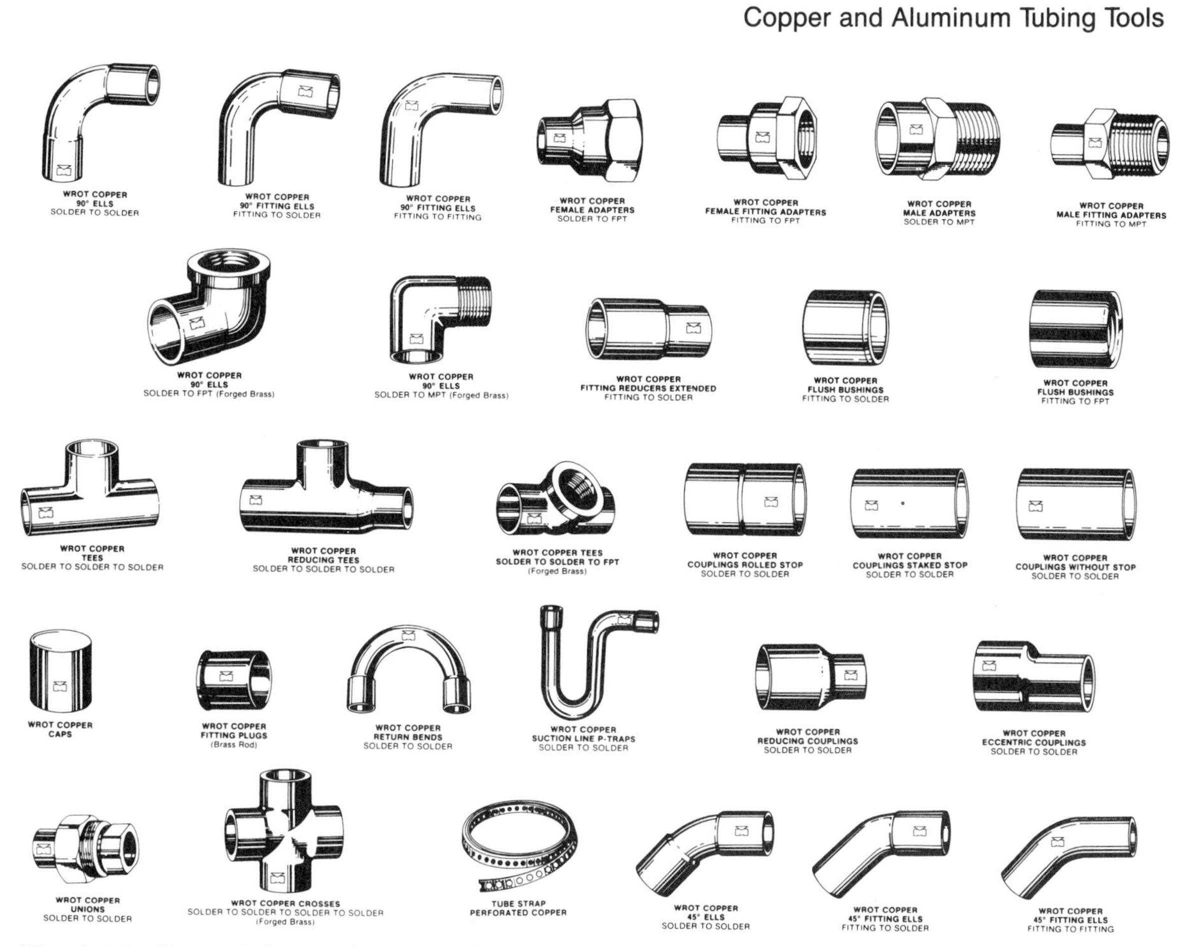

Fig. 2-20. *Copper fittings and adapters, including copper-to-copper and copper-to-iron pipe threads.* *Courtesy of Mueller Brass Co.*

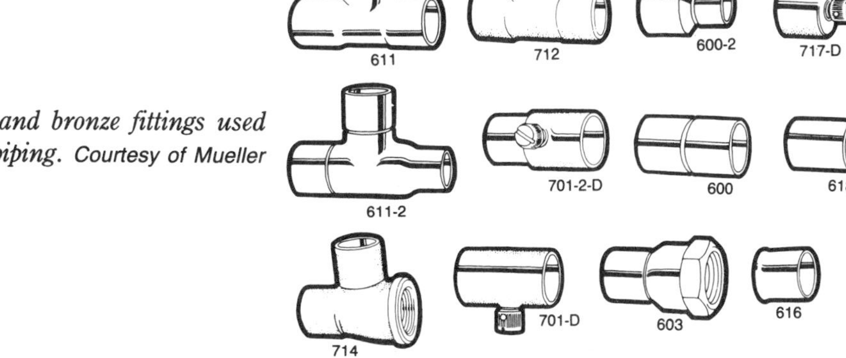

Fig. 2-21. *Copper and bronze fittings used especially for water piping.* *Courtesy of Mueller Brass Co.*

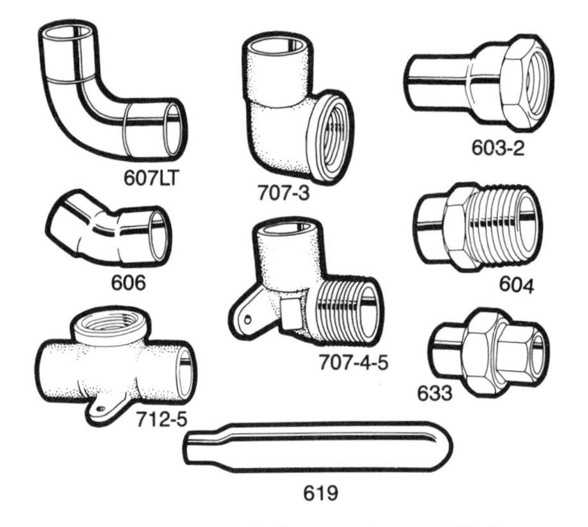

Fig. 2-22. Copper and bronze fittings used especially for water piping. Courtesy of Mueller Brass Co.

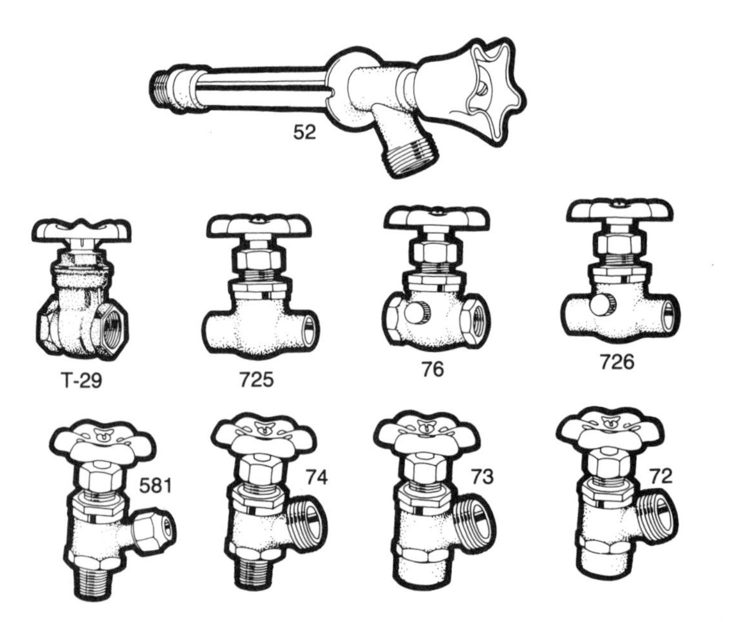

Fig. 2-23. Some stop cocks and special valves that are used for plumbing and heating. Courtesy of Mueller Brass Co.

3

Detecting Faulty Installations

BEFORE YOU BEGIN THIS CHAPTER, BE SURE THAT YOU READ CHAPTER 1 ON safety precautions. Working with heating and cooling systems can be dangerous if you do not take the proper safety precautions or know about fire hazards or other potentially dangerous situations. Keep safety in mind at all times.

OIL BURNING SYSTEMS

Many substandard, oil-burning furnace installation errors are obvious. Other hazards are less obvious:

1) The tank is too close to the burner, 10 feet is the minimum.
2) Absence of a fire extinguisher.
3) Improper mounting of the burner to the furnace or boiler.
4) Deterioration of the combustion chamber (in a conversion installation a separate combustion chamber is built or assembled of insulating fire-brick). Some combustion chambers are formed of stainless steel.
5) The smoke pipe from the furnace or boiler pitches downward toward the chimney instead of upward, and might not be sealed at the chimney with furnace cement.
6) There is no oil filter at the tank or burner.
7) Delayed ignition (burner starts with a bang).
8) Wrong size or spray pattern of the nozzle.
9) Improper pressure at the oil pump (normal is 100 pounds). See Figs. 5-1, 5-1A, and 3-1 through 3-8.

Fig. 3-1. A cut-away view of an oil burner nozzle that shows the internal parts, including the fine mesh screen. The nozzle is screwed into a nozzle holder and the holder is screwed onto a 1/8-inch oil pipe. Courtesy of Monarch Manufacturing Co.

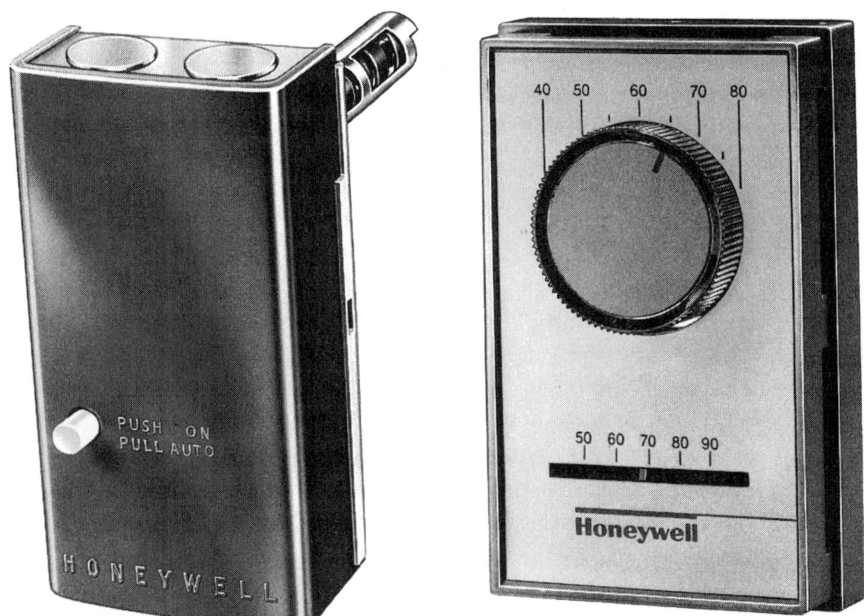

Fig. 3-2. A Standard heating thermometer. Courtesy of Honeywell Inc.

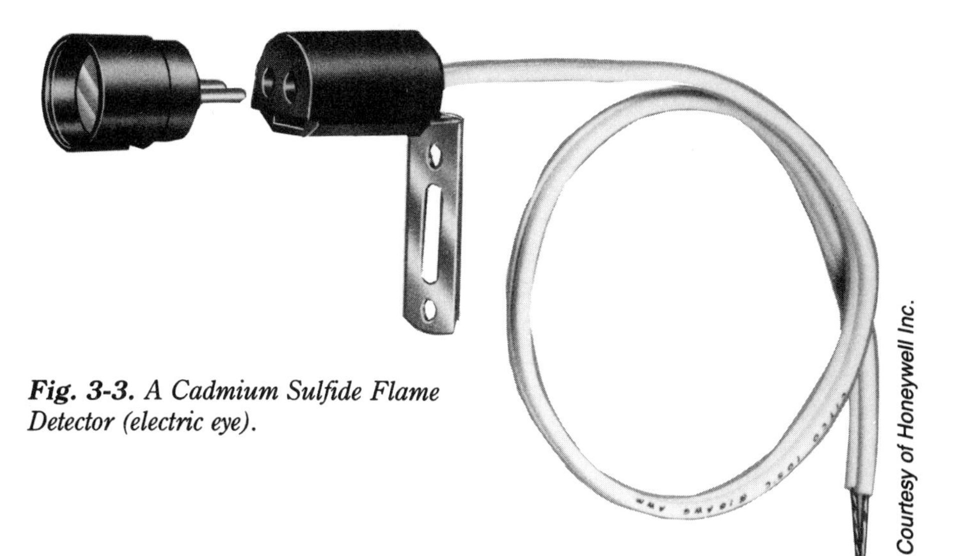

Courtesy of Honeywell Inc.

Fig. 3-3. *A Cadmium Sulfide Flame Detector (electric eye).*

Fig. 3-4. *A cut-away view of an oil burner nozzle. This nozzle has a "sintered" ball filter instead of a fine mesh screen.*

Courtesy of Monarch Manufacturing Co.

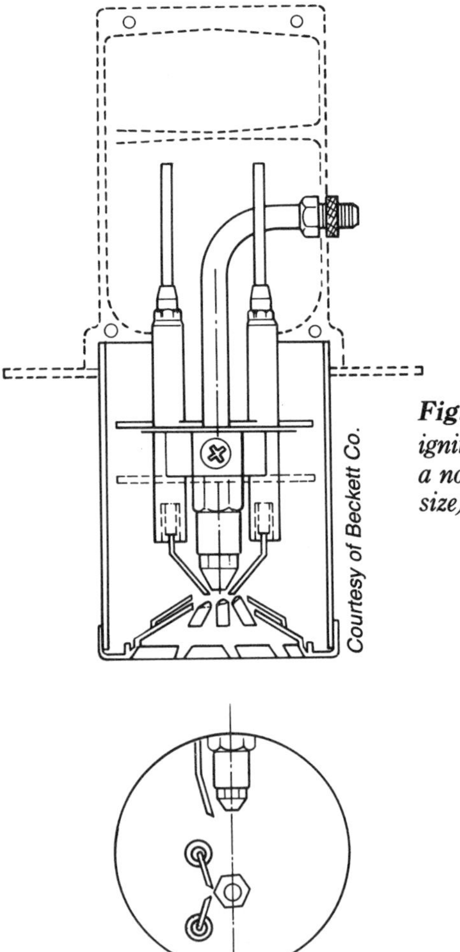

Fig. 3-5. *The assembled nozzle, oil tube, ignition electrodes, and flame retention of a nozzle and ignition electrode (quarter size).*

Oil Leaks and the Oil Line

A leak from a fuel oil line in a basement or utility room creates an unpleasant odor throughout the house. This is an important condition to check because it is a fire hazard, and any leaks must be repaired as soon as possible. Unfortunately, some systems have been installed where copper tubing is run directly on the floor and it is subject to crushing. This is poor policy and a hazard. Many times, the connections of the copper lines to the tank, pump, and nozzle are made using a type of fitting called compression. These fittings often leak and are *not* recommended for oil line connections. The best type of fittings are "flare" type, and are sold in most hardware stores. You will have to flare the end of the tubing, using a flaring tool. See Chapter 2 for instructions.

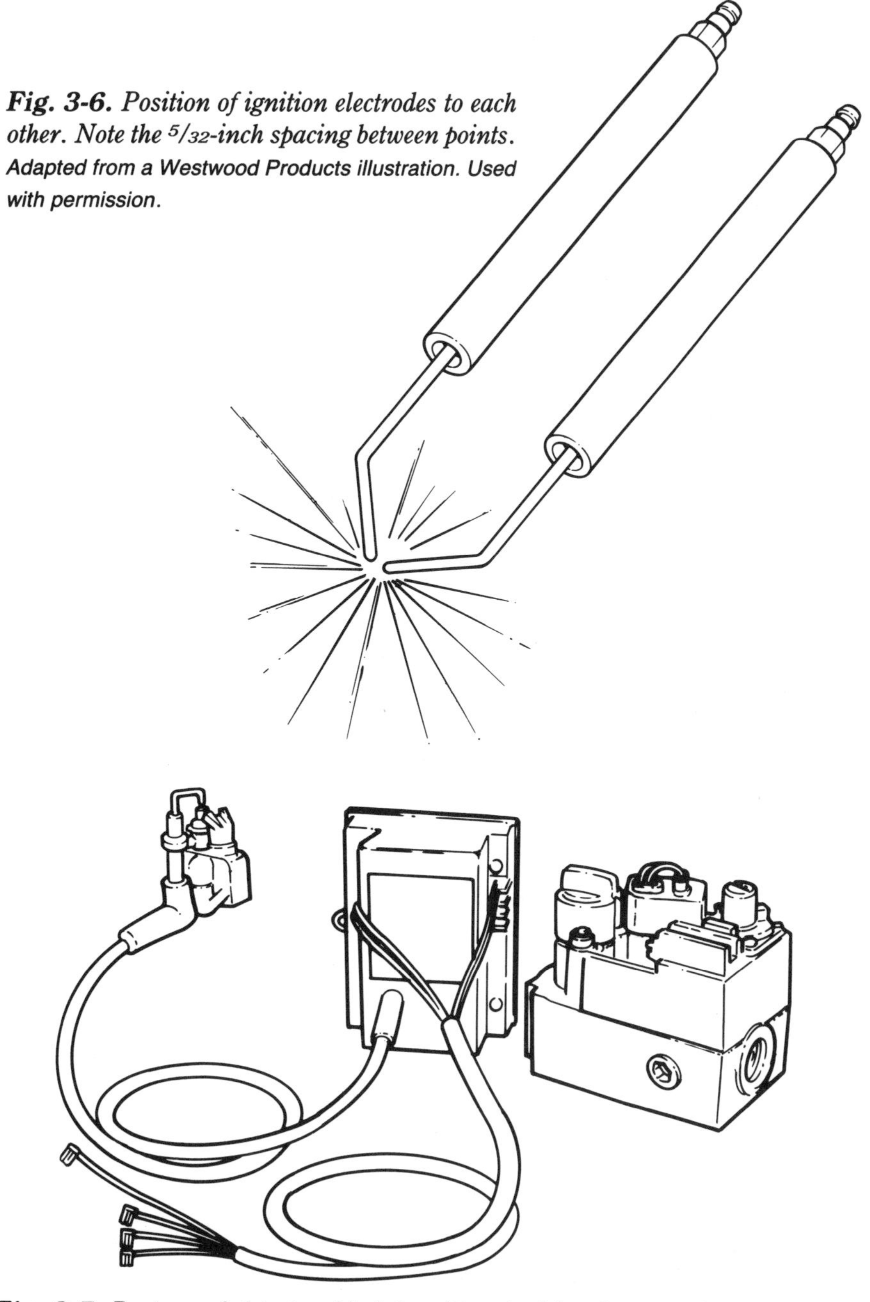

Fig. 3-6. *Position of ignition electrodes to each other. Note the $^{5}/_{32}$-inch spacing between points.* Adapted from a Westwood Products illustration. Used with permission.

Fig. 3-7. *Parts needed to provide intermittent ignition for a gas furnace: automatic gas valve, control module, pilot, and spark ignitor.* Adapted from a Honeywell photo.

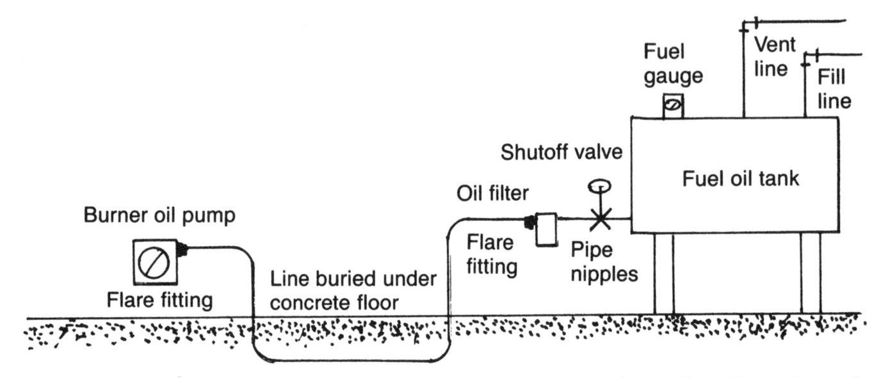

Fig. 3-8. *A diagram of an oil burner piping system, from the oil tank to the burner.*

Route the oil line from the tank to the burner so that no damage can occur to the tubing. Trench the concrete floor where foot traffic would likely crush the line, lay the tubing in the trench and cover with a mixture of cement and sand, troweled smoothly. Floor tile used in basements is easily damaged by fuel oil. Check for oil leaks occasionally, and repair or tighten the line when needed.

The piping arrangement of the oil line as shown in Fig. 3-8 is as follows:

1) The oil tank
2) The pipe nipple (and reducing bushing, if necessary to adapt the nipple to the tank tapping)
3) Shutoff valve
4) Nipple between valve and oil filter
5) Oil filter
6) Flare fitting, male pipe thread by $^3/_8$-inch male flare
7) Flare nut
8) Tubing to the burner location. The connection to the fuel oil pump will be $^1/_4$-inch male pipe by $^3/_8$-inch male flare, and a $^3/_8$-inch flare nut.

Controls

The older type of primary oil burner control includes a bi-metal helix coil that extends out from the back of the control about eight inches. The mounting flange is screwed to the smoke pipe and the "sensor" is inserted into the mounting flange. This "Stack Relay" (because it mounts in the stack) detects the presence of a flame in the combustion chamber because the smoke pipe becomes very hot. This causes the helix to twist; one end of the helix is fixed to the stack relay body, and the other end is attached to a system of contacts. These contacts make and break in sequence. This action "proves" the flame.

If the helix does not twist, certain contacts remain "made" and send current through a heating coil. After a delay of 30 to 90 seconds, this small heating coil warps another bi-metal that trips and shuts the burner down. In this case, the helix does not twist because *no flame* was detected. See Fig. 4-3. This is the safety shutdown (or lockout) initiated because of a component part failure of the burner assembly. Some relays permit a second "try" to start and light the oil, many do not, and the lockout mechanism has to be reset manually. This is the better method. **Note:** Do not try to reset the relay too soon because it must cool down after lockout. Wait five minutes.

The other type of relay for oil burners is mounted on the burner housing and uses an electric eye to "see" the flame. The eye is made of a cadmium sulfide material that is sensitive to the color of an oil burner flame (see Fig. 3-3).

Both types of oil-burner primary relays can be wired to provide constant or intermittent ignition. A burner wired for intermittent ignition operates as follows: On a call for heat, the burner starts and establishes a flame. When the burner starts, power is sent to the burner motor *and* to the ignition transformer. After about a two minute delay, the power to the ignition transformer is disconnected, stopping the spark. The flame continues to burn normally. Constant ignition means that the ignition spark continues as long as the burner operates. Under certain conditions, when the burner is wired for intermittent ignition, it can have problems. The terminal board on the relay has four wiring terminals marked: #1 - LINE, #2 - LINE, #3 - MOTOR, and #4 - IGN. For intermittent ignition, there are black wires on #3 and #4 terminals. To change to constant ignition, move the black wire from terminal #4 to terminal #3. Because only one wire is allowed under one screw, remove the wires from terminals #3 and #4. Using a short length of insulated black wire, form a loop on one end and attach it to terminal #3. Bare the other end of this short wire, straighten the bared ends of the original wires, then using a wire connector (wire nut), group all three bared ends and insert them in the wire connector. Tighten the connector firmly, and fold the wires neatly back inside the case of the relay. You now have constant ignition as the motor and the ignition transformer start and stop together. See Fig. 4-3. Most modern oil burners are designed with a "flame retention" head, which keeps the flame closer to the nozzle at all times (see Fig. 4-1).

An important adjustment to correctly make is the air to fuel ratio. Before disassembling a burner, any adjustments, settings, or position of parts should be marked so you can return them to their original positions. This advice holds true for any settings that must be changed temporarily and later returned to the original setting.

The oil burner body has an adjustable band on the air intake of the burner (the side holding the oil pump). This band could accidentally be moved when working on the burner. A pencil mark or scratch will identify the original location of the band. Excess air, caused when the air intake band is open too wide, might not allow the burner to light. If the burner does light, the flame will be white and emit an acrid odor. A normal flame is orange with slightly smoky tips. A very smoky flame indicates too little air. Other adjustments you should check and change as necessary include:

1) The oil pressure at the nozzle. A normal setting is 100 pounds per square inch (PSI).
2) The carbon dioxide CO_2 reading. A normal level is 12 1/2 percent to 13 1/2 percent.
3) The draft—normal .01 inches to .02 inches water column (W.C.).
4) A smoke test. Refer to Fig. 3-5 for adjustment details.

GAS BURNING SYSTEMS

Similar errors in an oil burner installation are found in a gas burner system. If the gas line is not supported properly, joints can leak. A dirt leg might not have been installed. Instead of using an elbow to turn the vertical gas line into the furnace, a tee is used. Then a four to six-inch nipple is screwed into the tee and a pipe cap is screwed onto the lower end of the nipple, see Fig. 4-12. This forms a pocket for chips and dirt particles to collect. A shutoff valve and union might not have been installed ahead of (upstream of) the furnace automatic gas valve. Without the valve and union, the furnace cannot be disconnected from the incoming gas line without cutting the line or disconnecting the line at the meter. This is a code violation and, if so installed, the furnace was installed without a permit or inspection.

Other grave errors of this type might not be so obvious, and it is important that the city inspector be called. If his recommendations are followed, the installation is certain to be safe. Many municipalities now require that a house be inspected before being sold to another person.

The Thermocouple

Occasionally the thermocouple fails, either it is defective, or the end of the tube that is screwed into the automatic gas valve is loose. If this is the case, tighten the connection using a small wrench. CAUTION: As this is an electrical connection, do not tighten it excessively, only snugly. If you must replace the thermocouple, they are available at most hardware stores. Match the length as close as possible. Also determine that the pilot end will fit in the bracket on the pilot. Usually the thermocouple can be removed from the pilot bracket. Sometimes the complete pilot and thermocouple must be removed as a unit, or the burner assembly, which holds the pilot bracket, must be removed as a unit. Study the area and decide the best way to remove the assembly. First, shut the gas off at the shutoff valve outside the furnace casing. Disconnect the pilot line and thermocouple lead. The burner itself can be removed by pushing the burner assembly toward the rear of the area. This removes the burner from the gas orifices on the manifold. The burner can now be lifted out, pilot, thermocouple, burner, and related tubing as a unit. Each furnace is slightly different, and you might need to study it some before removing the burner.

An older house might need extensive work done to meet the requirements of newer building codes, but the new buyer is assured of a safe house, mechanically and structurally.

FORCED, WARM-AIR SYSTEMS

Before evaluating how a forced, warm-air furnace produces heat, it is a good idea to begin with the furnace itself. Start by examining the outside of the furnace casing, check for bent, missing, or damaged panels. At the rear of the furnace is a removable panel, this panel permits access to the air filter(s), blower, and blower motor. Remove the panel, then check the filter and the filter frame.

Air Filters

The filter should fit the frame snugly so that all of the air passes through the filter. This is especially important if an air conditioning coil is installed in the plenum of the furnace. If air can bypass the filter, the air conditioning coil might become clogged with dirt. Because the coil is built into the plenum, it will be difficult to clean. Some furnaces require a special size filter, and a standard size filter that does not fit properly might have been substituted. Most hardware stores and home centers now stock cut-to-fit filters that can be used. The local dealer of that make of furnace might also have filters to fit.

If the present filter fits, the thickness should not be more than one inch. Two-inch filters cut down on airflow, and the furnace fan is not designed to operate with the greater pressure drop of the two-inch filter.

The Fan and Fan Belt

Inspect the blower (fan) by holding a light so you can see the condition of the blades. Scrape one blade to check for dirt buildup. If enough dirt is on the blades, the blower cannot circulate enough air through the furnace and the limit control will shut down the burner. On an older furnace, you might have to remove the blower and carry it outside for a good cleaning.

If the blower is belt driven, check the condition of the belt. The belt should be tight, but not too tight. Press down midway between the pulleys. A one-half- to three-quarter-inch slack is correct. The belt should not have any cracks, especially on the inside. A very shiny belt indicates that the belt is slipping. The pulley on the motor might be the "adjustable" type. This pulley has two halves that screw together or apart, to change the effective diameter of the pulley. When the pulley halves are screwed together, making the pulley larger, the blower will run faster, moving more air through the furnace. The blower and motor generally have oiling ports. Using **nondetergent electric motor oil** only, lubricate the blower and motor sparingly. Use only three to five drops twice a year on the motor. Most hardware stores and home centers sell this type of oil in small cans labeled "electric motor oil." Do **not** use automobile engine oil!

The Combustion Chamber

Inspect the interior of the combustion chamber using a mirror and trouble light or a flashlight. The walls should be fairly clean if the burner has been maintained regularly. If there is much soot, you can remove it using a shop vacuum with a brush adapter, or rent similar equipment. After a number of years, the flue pipe might need replacing. It is very important that the joints of the flue pipe be tight with no leaks. Combustion products inside a house are a very dangerous condition. Carbon monoxide, a product of combustion, is a deadly, odorless, poisonous gas and is present anywhere there is combustion!

All joints of the flue pipe must be as airtight as possible to prevent the escape of any gases into the house. The opening in the chimney where the flue pipe is connected must be tightly sealed with furnace cement or boiler putty. Furnace cement is sold at all hardware stores. The chimney should have a "clean-out" at its base. Open the clean-out door, and using a small mirror, inspect the inside of the chimney for any soot or other debris. If aimed properly, daylight or the sky can be seen in the mirror. Light can usually be seen, even in offset chimneys. A chimney can be cleaned by working a chain up and down, while standing on the roof. It might be a good idea to hire a chimney sweep to do this cleaning. Chimney sweeps are listed in the telephone book.

Controls

All types of heating equipment have safety and operating controls. A fan control is needed to operate the blower (fan). This control is generally set to turn the fan on at 130°F and off at 90°F. This prevents the fan from circulating cold or cool air at any time. A furnace limit control, sometimes called a "high limit," is necessary to turn the burner off if for some reason air is not circulated through the furnace. The control will stop the burner when the plenum temperature rises to 200°F. See Fig. 3-2. Usually this control must be manually reset so that someone will check to determine the cause of the problem that caused the limit control to operate. Excessive plenum temperature can be caused by one of the following:

1) A broken fan belt
2) The fan motor does not start
3) The fan wheel is very dirty
4) The filters are very dirty (plugged), or the cooling coil is plugged.

A forced, warm-air furnace cycle begins with the thermostat. The thermostat calls for heat, the burner starts, the furnace plenum temperature rises to the On setting of the fan control. The fan starts and continues to run, the thermostat becomes satisfied and shuts the burner Off, the fan continues to run until the plenum temperature is reduced to the Off setting of the fan control, 90° or 95°F. The fan then stops until the next call for heat. The frequency of these cycles varies with the outdoor temperature. The colder the outside, the more frequent the On cycles. At 0°F, the furnace will run constantly, as this is the design temperature for the furnace, and if the furnace is sized correctly for the house and the amount of insulation, this constant operation is normal.

Electrical Wiring

The furnace or boiler's electric wiring should be inspected. It is important that the wiring connections be tight. Loose connections overheat and contribute further to the looseness. Usually, terminal screws can be tightened one whole turn, sometimes two turns. Low voltage from loose connections can affect the operation of motors and controls and perhaps damage them. Check device enclosures for excessive heat by touching them. If an enclosure feels hot, shut off the power and check for loose connections. Circuits fused for 15 amps will protect a number 14 copper wire; circuits fused for 20 amps will protect a number 12 copper wire. Fuses larger than these should not be used, and each size fuse should be used to protect its correct size wire.

Fuel Leaks and the Fuel Line

Fuel oil lines that are connected with compression fittings are subject to leakage. All oil line connections are now required to be made using flare nuts and fittings. These flare fittings seldom leak when made properly. Oil lines not supported or fastened in place can be damaged, and might leak at points along the oil line between mechanical connections.

It is wise to check for leaks in the gas piping to a gas burner. A quick way to check is to sniff at each threaded joint. All gas has an additive added for leak detection that can be smelled, and has a strong odor. If you smell gas, recheck using a thick soap solution or a commercial solution. A leak will form a bubble. Any leaks must be repaired *immediately*, as an explosion could result if the leak is not repaired. Any mechanical device is subject to malfunction or failure, therefore all such equipment should be checked periodically. Be alert to the noise of your system's normal operation so that you can detect any extra or different noise and take steps to determine its origin. For example, a fan belt that squeaks when the fan starts, or the burner makes a loud noise when it lights. Repairing faulty equipment often saves time and money, not to mention the inconvenience of no heat or cooling.

4

Evaluating Gas- and Oil-Fired Furnaces

THE BEST TIME TO INSPECT YOUR SYSTEM IS DURING THE SUMMER OR DURING A period of mild weather so that the furnace can be turned off for two or three hours. Any system, whether gas or oil fired, needs to be evaluated. This evaluation should include the overall appearance and any obvious errors of a poor or faulty installation, but before you can effectively evaluate your system, you need to know a little about how each works.

THE OIL BURNER

The oil burner differs from the gas burner in many of its component parts, but both types of burners have a controlled flame in the combustion chamber. The oil burner is controlled by three devices:

1) The thermostat
2) The primary control (stack control or electric eye)
3) The safety controls

An oil-fired furnace begins to work when the room temperature drops below the thermostat setting, the thermostat *makes* contact, sending a signal to the primary control. The primary control contacts close, starting the burner motor and energizing the ignition transformer. When the burner starts, the oil pump on the burner sprays oil into the furnace combustion chamber. A spark from the ignition transformer ignites the oil spray and establishes a flame and the burner continues to run.

The Primary Control

The primary control and the safety control can be housed together in one case or separated, as is the electric eye which must "see" the flame and is mounted in the fire door or inside the blast tube of the burner. The older type of safety control is located in the smoke pipe and senses the presence of a flame by the rise in temperature.

A spark from the ignition transformer should ignite the oil spray and establish a flame. If a flame is not established after a predetermined number of seconds, the primary control shuts the burner down. Most primary controls will lock out under these conditions. Some primary controls will make one more attempt to start the burner and establish a flame before locking out. The time delay varies with the manufacturer of the furnace and is specified by them. Usual delays vary from 15, 30, to 45 seconds. Refer to Figs. 4-1, 4-2, and 4-3.

Controls found on an oil-fired forced warm-air furnace are:

1) The Primary control, either a single relay mounted in the stack (smoke pipe), or a two-piece relay consisting of a Primary relay and an electric eye to "see" the flame.
2) The Fan control. This control starts the fan when the burner raises the furnace temperature to 130°F. After the burner stops, this control allows the fan to run until all of the heat is exhausted from the furnace and stops the fan at about 90°F. Both of these temperature settings are adjustable within the range of the control.
3) The Limit control, often combined with the fan control in a single case, is a safely control to prevent the furnace from becoming overheated (see Fig. 4-2). Overheating is caused by the fan not running due to a broken fan belt, a dirty centrifugal switch in the fan motor, frozen bearings in either the fan or fan motor, or very dirty air filters. If the furnace temperature rises to 200°F, the limit switch will stop the burner. After the furnace has cooled down to about 160°F, the burner is allowed to restart. This differential, which is the difference between Off and On of a control, is set at the factory; a stop also prevents the limit setting from being raised above 200°F.

Fig. 4-1. *A modern oil burner. The flame retention ring can be seen in front of the blast tube, the motor is on the left, the oil pump is on the right, and the ignition transformer on top.* *Courtesy of Beckett Manufacturing.*

Fig. 4-2. *An oil burner primary safety relay used with the Cadmium Sulfide Flame Detector (electric eye).* *Courtesy of Honeywell, Inc.*

Fig. 4-3. *An oil burner primary safety relay, such as this stack mounted model, reacts to heat in the smoke pipe—or stack.* Courtesy of Honeywell, Inc.

Boiler Controls for Hot-Water and Steam

Controls found on an oil-fired boiler perform functions similar to those on a warm air furnace. The burner control is the same. An Operating control maintains a set temperature of the boiler water, on modern systems at 200°F (older systems may operate at 175°F). The high-limit control is set at 210°F. A low-water control is to prevent the boiler running out of water (running dry) and consists of a float switch. If the water level drops, the float will open the circuit and stop the burner. This control was originally designed to be used only on steam boilers, to prevent a low-water condition which is more critical to steam boilers than hot-water boilers. Later it was found that hot-water boilers drained for repairs or cleaning were not locked out electrically. Persons would start the burner operating, not knowing that the boiler had been drained. In many cases, the boiler was severely damaged. See Figs. 4-4 through 4-7.

Newer, hot-water systems are designed so that the thermostat controls the hot-water circulating pump, which runs only when the building temperature falls and the thermostat calls for heat. As stated above, the boiler maintains its water temperature at 200°F, and the pump circulates water to the building as needed.

Steam boilers are controlled directly by the thermostat that operates the burner. An operating control maintains a steam pressure of two to five pounds per square inch (PSI). The high limit control is set at nine PSI. A low-water float operated control shuts the boiler off if the water level is low enough to cause excessive damage to the boiler; this is required by the Boiler Code. In addition, a Safety Relief Valve set to relieve at 15 PSI is also installed to protect the boiler.

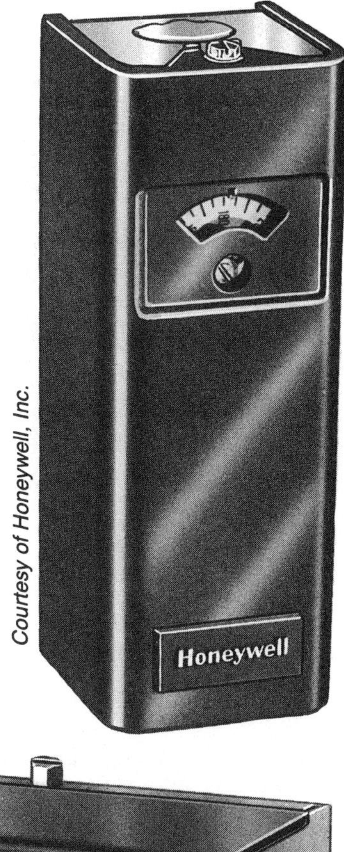

Courtesy of Honeywell, Inc.

Fig. 4-4. *A limit control maintains the boiler water temperature.*

Courtesy of Honeywell Inc.

Fig. 4-5. *An acquastat controls the boiler temperature and does not let the circulating pump run if the boiler water is too cold.*

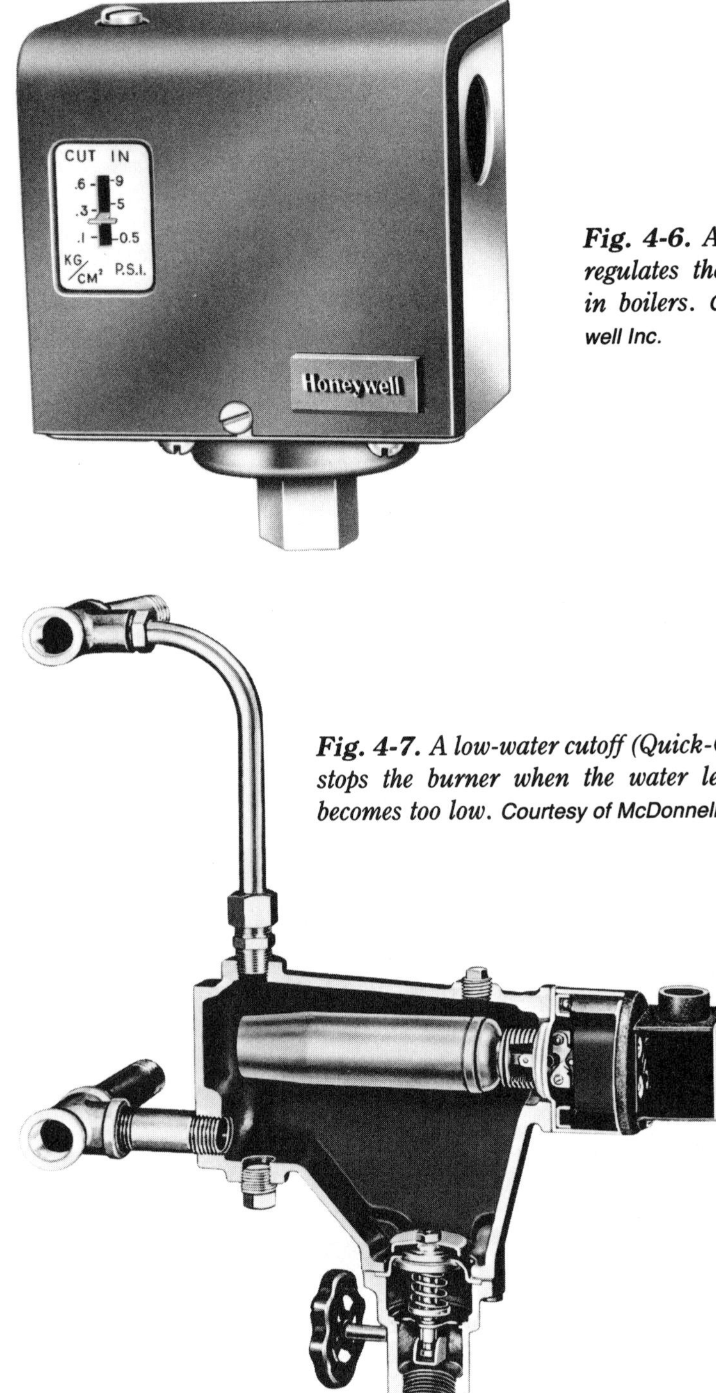

Fig. 4-6. *A pressure control regulates the steam pressure in boilers.* Courtesy of Honeywell Inc.

Fig. 4-7. *A low-water cutoff (Quick-Connect type) that stops the burner when the water level in the boiler becomes too low.* Courtesy of McDonnell and Miller Co.

The Oil Burner Mechanism

Most modern oil burners are "pressure type," meaning that the oil is forced, under 100 pounds per square inch (PSI), through a very small orifice. This forms a cone of oil droplets. This oil spray is ignited by an electric spark blown into the cone as shown in Fig. 11-22.

The burner consists of a housing, usually cast aluminum, that holds the burner motor, a squirrel cage fan, a fuel pump, an ignition transformer, two ignition electrodes, a nozzle, and an electrode holder. Refer to Fig. 11-14.

The motor faces the rear of the burner on one side and the fuel oil pump and the adjustable air intake are on the opposite side. The fan is between the motor and oil pump. On the motor shaft is the fan and a flexible coupling that connects to the oil pump.

Below, and accessible through a rear cover, is the electrode and nozzle assembly. Flexible wires or spring contacts connect the rear ends of the electrodes to the ignition transformer. The electrodes are mounted on the oil pipe, which supplies oil to the nozzle.

At the rear near the access cover, the oil line turns and leaves the "blast tube" on the side where the oil pump is located so that it can be connected to the oil pump. The front of the oil pipe has an adapter to hold the nozzle. A bracket clamped around this pipe holds the electrodes and also centers the pipe and nozzle in the blast tube. Figure 3-7 illustrates this assembly.

The two electrodes look somewhat like a spark plug and supply a spark to ignite the oil spray. Each electrode consists of a long (four-inch), porcelain tube about 9/16 inch in diameter. A small hole through the center holds the actual electrode wire. The rear end of the wire is threaded for a nut to hold it in position and connects it to the ignition transformer. The front end of the electrode wire is bent close enough for a spark to jump to the other electrode. This spark "gap" is usually about 5/32 inch, but varies depending on the manufacturer. The two electrodes are positioned *behind* and out of the oil spray as the force of the air from the fan blows the spark into the spray and ignites the oil as shown in Fig. 11-22.

Fuel Supply and Storage

The fuel oil tank for a single family house is usually located in the basement, or in some cases, outside near the rear of the house. The fuel tank and fuel oil pump are connected with copper tubing, either 3/8- or 1/2-inch outside diameter (OD). Inside tanks must be located a minimum of 10 feet from the furnace. Many fire codes allow two oil tanks in a basement, provided there is a three-way valve between the tanks. Consequently, only *one* tank can be connected to the burner at any one time.

The fuel line contains a fuel oil filter to remove any particles that might clog the oil nozzle. This filter should be either at the tank itself or at the fuel oil pump on the burner. A shutoff valve is mounted directly on the fuel tank so that the filter can be changed and the burner serviced when necessary.

For larger homes, the oil tank is sometimes buried underground to provide a larger oil reserve. Buried tanks are subject to water seeping into them. This occurs because the tank might have settled, and a welded joint opened, allowing water to enter. Water in the oil will stop the burner or cause it to operate improperly. An underground oil tank is also piped differently than an inside tank. An underground oil tank uses a ''two-pipe'' system.

Nearly all fuel oil pumps are designed for either a one- or two-pipe system. Plugged openings are provided for setting the pump for use on either system. If a tank is used for two pipes, a plug is inserted inside the pump body to close off a bypass port. These two-pipe systems are not often found today. The main difference you will find is two pipes between the oil pump and the oil tank. Both lines enter the tank at the top, the suction line dips to 12 inches above the bottom and the return line does not dip down but only enters a tapped fitting in the top of the tank. All fuel oil pumps have tapped openings marked ''suction'' and ''return.'' Printed directions are usually furnished and illustrate how to insert the ''bypass'' plug using an Allen wrench. Refer to Fig. 6-5 and 6-7.

SERVICE AND MAINTENANCE

Properly serviced and maintained, the modern oil burner is a clean and reliable source of heat for the homeowner. The burner and related controls are very easy for the average handyperson to repair and maintain. Repair parts are available at some hardware and home center stores. Specialized parts can be found at heating suppliers and motor shops that sell to the individual. If you are going to replace a part, always take the defective part with you so that the clerk can supply the correct part. Oil burner motors differ from the standard motor, which is held in a cradlelike base. The burner motor is flange-mounted. The motor has the flange on the shaft end with two holes for bolts (sometimes three) in the flange. Different makes of burners usually have different size flanges and bolt-hole positions on their motors.

The Centrifugal Switch

Oil burner motors and furnace fan motors have a small centrifugal switch. This switch is housed on the end-bell where the electrical wires enter. Frequent oiling and dirt being drawn inside cause the contact points to become dirty. These points do not easily become pitted, only dirty. It takes about a half hour to remove the end-bell, clean the points, and replace the end-bell.

To clean the contact points, use a business card and draw the card through and between the closed points. The card will be streaked with black oily dirt. Keep doing this until the card comes out clean. The symptom of dirty points is that the motor will not start but will hum while trying to start. Eventually the motor will trip its overload device or blow a fuse or circuit breaker.

To clean the centrifugal switch, first shut off the furnace at the fuse or breaker panel. Remove the motor from the burner after disconnecting the wires from the motor. Tape the wire ends for safety. Before removing the end-bell, mark the motor body and end-bell with a prick punch to make it easy to realign the two parts. There will be four through bolts holding both end-bells to the body. Remove these bolts, then gently tap with a cold chisel or screwdriver around the perimeter of the end-bell to be removed. Part of the rim of the end bell is raised and you can start tapping at these points. Continue tapping equally around the perimeter, being careful not to tilt the end-bell. Some end-bells release easily; some do not. Have patience and do not force the end-bell. The centrifugal switch is mounted on the inside of the end-bell. Wires will lead from the switch terminals to the windings of the motor, so do not put a strain on these wires. If the switch mechanism is badly worn or some part is broken, you'd be better off to buy a new switch. You should be able to find a switch at appliance repair parts centers or motor repair shops. The wires connected to the switch may be quite short. The points of the switch can be easily seen. As noted above, draw a business card between the closed points to clean them. Do not file or sand the points. Because of over oiling, the burner motor and the fan motor should be oiled sparingly. Use three to five drops of *non-detergent electric* motor oil in each oil port twice a year. By doing this, you might prolong the time when the centrifugal switch needs to be cleaned. Figures 4-8, 4-9, and 4-10 show these procedures.

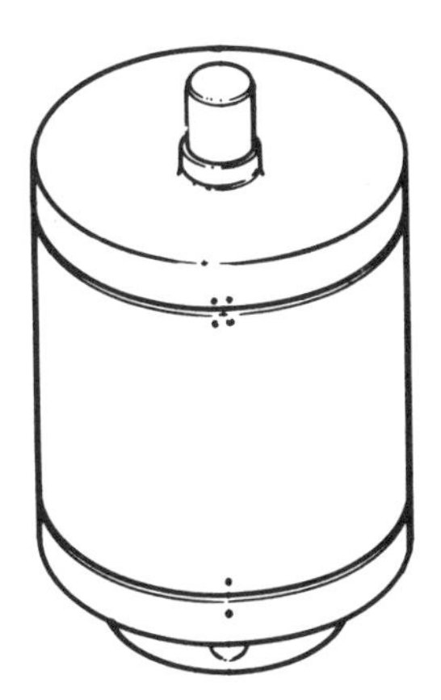

Fig. 4-8. *A motor with an internal switch, ready for disassembly. Punch marks were made on the endbells and directly opposite on the center housing.*

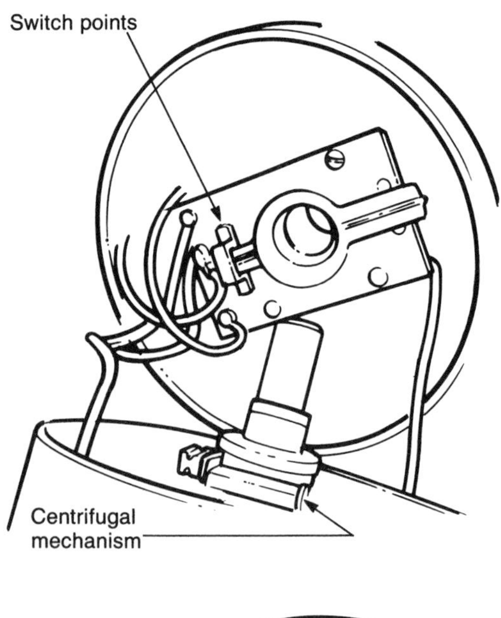

Fig. 4-9. The endbell holding the centrifugal switch has been removed. The arrow indicates the switch contacts. The centrifugal mechanism is still on shaft (bottom).

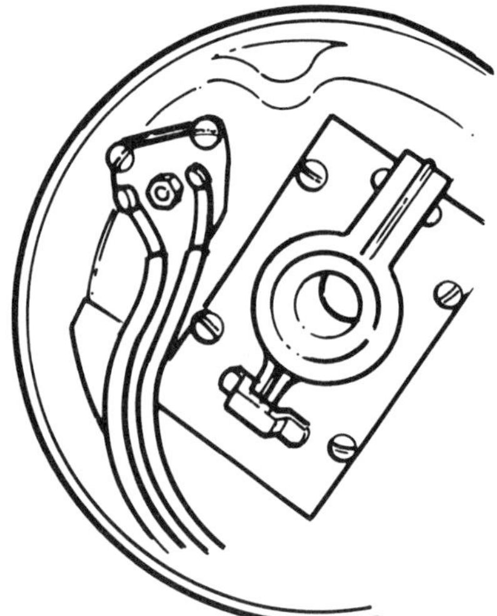

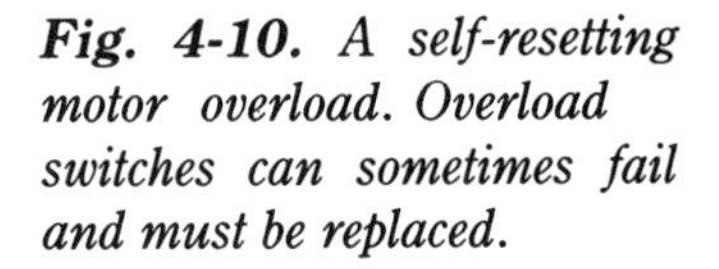

Fig. 4-10. A self-resetting motor overload. Overload switches can sometimes fail and must be replaced.

THE GAS BURNER

The gas-burner and oil-burner are similar in that both provide heat to a heat exchanger, whether forced, warm air or a boiler. The safety controls for a gas burner are entirely different from those of an oil burner, however. The standard design is a gas burner/warm air furnace combination, designed as a package with both parts used together as a unit. Many years ago, when natural gas was first made available for home heating, conversion gas burners were manufactured to be installed in coal-fired furnaces and boilers. This was a great improvement over the use of coal. The modern, gas-designed forced-air furnace has similar controls.

Because the warm-air furnace or boiler using a gas conversion burner is still in use in many homes, directions for servicing and repairing these burners are covered in this book. While the burner gives good service, it is not as efficient as newer, gas-designed furnaces or boilers. First, as always, the safety controls are the main controlling device. These three devices in combination are:

1) The pilot burner, which has two functions: to light the main burner when required and to heat the thermocouple, see Fig. 11-12.
2) The "Baso" pilot switch is a small box on the front of the furnace having two electrical contacts. The other end of the thermocouple line is connected to this Baso switch.
3) The main automatic gas valve.

The Gas Line

A gas heating system starts at the gas meter, either in the basement or outside mounted on the house wall. Black iron pipe about 1- to 3/4-inch size comes from the gas meter to the furnace. This part of the gas line might be one inch and reduced at the furnace. At this point, it might branch off using a pipe tee to also feed the hot-water heater, if the heater uses gas. The gas line is run on the underside of the floor joists (or between them if the direction is right), and attached to the joists. The line drops down beside the furnace using an elbow (el) or a tee, if the line goes on to the water heater. About five feet above the floor is a shutoff valve. The line then continues down to the level of the burner main, automatic valve, see Figs. 4-11 and 4-12. Instead of the line turning to meet the valve, a tee is installed, then a short pipe nipple pointing to the floor. This open end is closed using a pipe cap. This is called a "pipe trap" to trap sediment and metal chips from the pipeline. The side opening of the tee points to the automatic valve and is connected to that valve. A pipe union is needed to make the final connection. Note furnace switch.

The shutoff cock at the five foot level has a small shutoff valve called a "B-Valve" screwed into the larger valve on the side of the valve towards the gas meter. The B-valve feeds the furnace pilot burner. This arrangement allows the larger valve to be shut off while the small valve can be opened to allow gas to supply the pilot valve only, then the pilot can be lighted independently with no danger of the main burner lighting while trying to light the pilot. The automatic

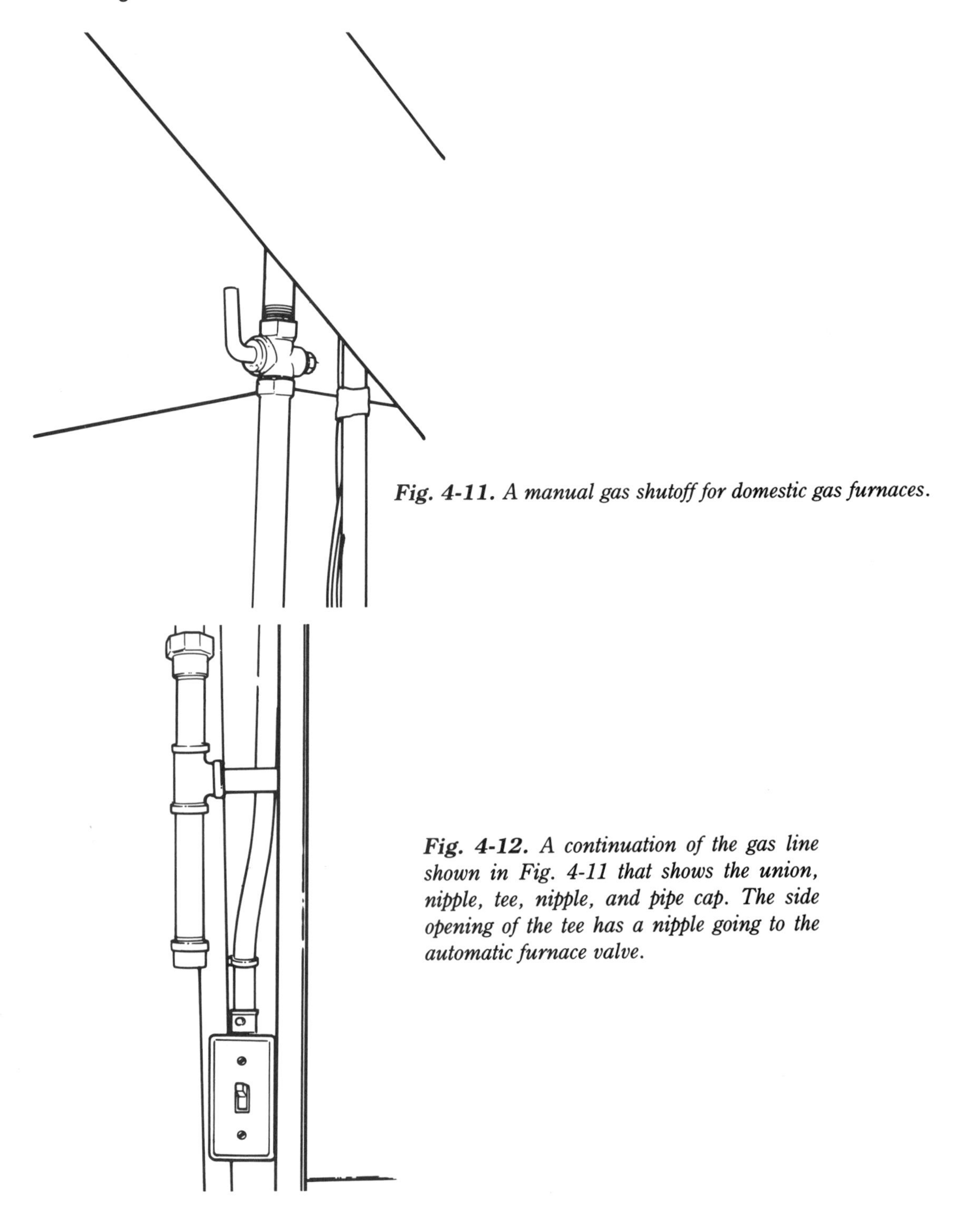

Fig. 4-11. A manual gas shutoff for domestic gas furnaces.

Fig. 4-12. A continuation of the gas line shown in Fig. 4-11 that shows the union, nipple, tee, nipple, and pipe cap. The side opening of the tee has a nipple going to the automatic furnace valve.

valve is usually a solenoid; newer burners may have a diaphragm type as these are quieter than the solenoid type. After leaving the automatic valve, the gas line continues on to the burner. Some burners have a manifold with three or more burner heads; other burners called ''fountain'' type feature an undershot tube that curves up to an upright position. The gas flame shoots out up against a round baffle that spreads the flame in a horizontal fountain effect for better heat distribution in the fire pot.

The Thermocouple

The thermocouple is attached to, and heated by, the ''standing pilot'' (so-called because the pilot is always burning, rather than intermittent). Most pilots have three fingers of flame, two to light the main burner, and one to heat the thermocouple. A thermocouple consists of two dissimilar metals joined at one end. When a voltmeter is connected across the other *separated* ends, a very small voltage is detected, 50 millivolts (1 volt equals 1,000 millivolts). At the end of the thermocouple opposite the end being heated by the standing pilot, is an electrical connection, which fits a mating connection on the Baso switch. This switch has a small solenoid as a part of an electrical switch. Because the small voltage is not sufficient to lift the solenoid, a reset lever is pressed, raising the solenoid and closing the electrical switch. This action allows current to energize the automatic valve, and the valve will open, provided that the thermostat is calling for heat. The safety feature of this system is; if the standing pilot should go out for any reason, the Baso switch will open because the thermocouple does not generate current and thus breaks electrical contact at the Baso.

5

Selecting a Heating System

CHOOSING THE RIGHT TYPE OF HEATING AND COOLING EQUIPMENT IS important to providing satisfactory operation. Equipment that performs its function with little or no noise, maintains the set temperature, can be relied upon to operate when required, require a minimum of maintenance, and be economically feasible. To achieve these goals, the equipment selected should be manufactured by a reliable company. Name brands made by national manufacturers can usually be depended on to give satisfactory service over their lifetime. Periodic maintenance you must do once a year includes oiling bearings (burner and blower motors), cleaning, adjusting, and checking the component parts of the burner, boiler, or furnace. Some gas furnaces or boilers need this work done only every *two* years, however, I suggest that an evaluation be made *every* year. You don't necessarily need to dismantle the components. Use a flashlight and check the system thoroughly, check the fan belt for cracks and wear, change the air filters twice a year, and oil the bearings, sparingly.

The homeowner is faced with a vast array of new, innovative equipment for both heating and air conditioning. To help you to better understand it, I've divided heating equipment into two types: forced warm-air and hot-water heating (also called "hydronic" or "wet heat"). **Note**: The following section covers *only* the furnace or boiler and does not discuss the type of heat source.

FORCED, WARM-AIR HEATERS

A forced, warm-air heater is the most prevalent type of system because it is fairly easy to install yourself (except for the ductwork, which can be contracted for through a local sheet metal shop). The particular advantage that this type of heating system has over other types is the ease of adding central air conditioning to the system, during installation or at a later time. The warm-air system has a quick response or recovery, such as when the thermostat is set back at night, to provide heat quickly when needed. It is a relatively quiet system, although air movement through the ductwork and registers might make some noticeable noise. A larger blower (fan) running at a slower speed can help to reduce air noise if necessary.

Air conditioning can easily be added later, even after the heating system has been installed. A special cabinet is available from the furnace manufacturer to house the cooling coil. This section sits atop the furnace and the ductwork is connected to this section. NOTE: The ductwork must be connected to this section using a flexible connector made of fire resistant fabric. This prevents any vibration of the furnace from being transmitted to the ductwork and to the rooms above. If you anticipate installing cooling at a later date, you can include this cabinet when the new furnace is installed at a small additional cost. You can also get a three-speed furnace blower (fan) to be used later with a cooling system. This way, you need only install the air conditioner's outdoor condensing unit and the furnace's cooling coil to provide air conditioning to the whole house. If you are replacing an old furnace, add a humidifier, or one can be added at a later date if this is new construction, because basement walls and concrete floors usually have only enough moisture in them to compensate for lack of humidity during the first year of occupancy. Refer to Figs. 5-1 through 5-4.

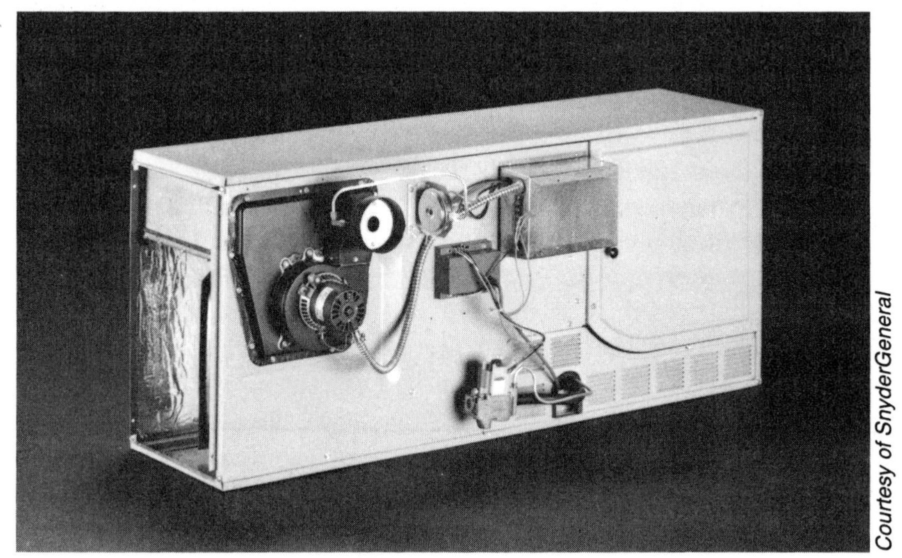

Courtesy of SnyderGeneral

Fig. 5-1. *A gas-fired, induced-draft, horizontal furnace.*

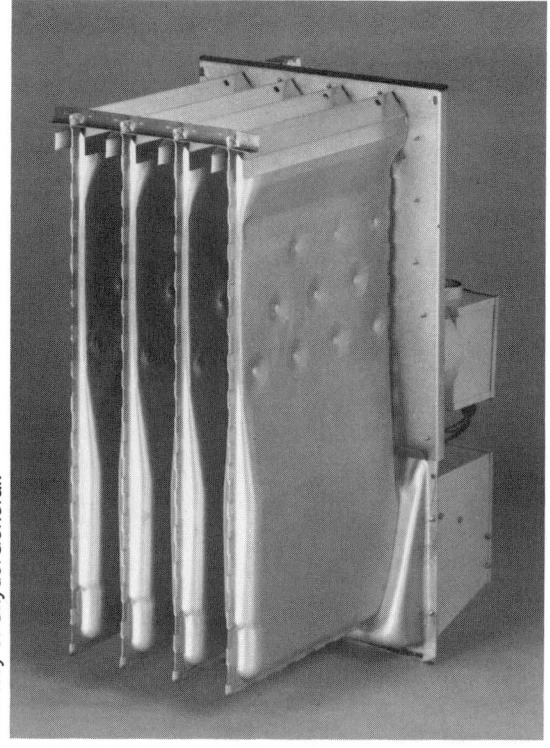
Courtesy of SnyderGeneral.

Fig. 5-2. *A heat exchanger for a gas-fired furnace.*

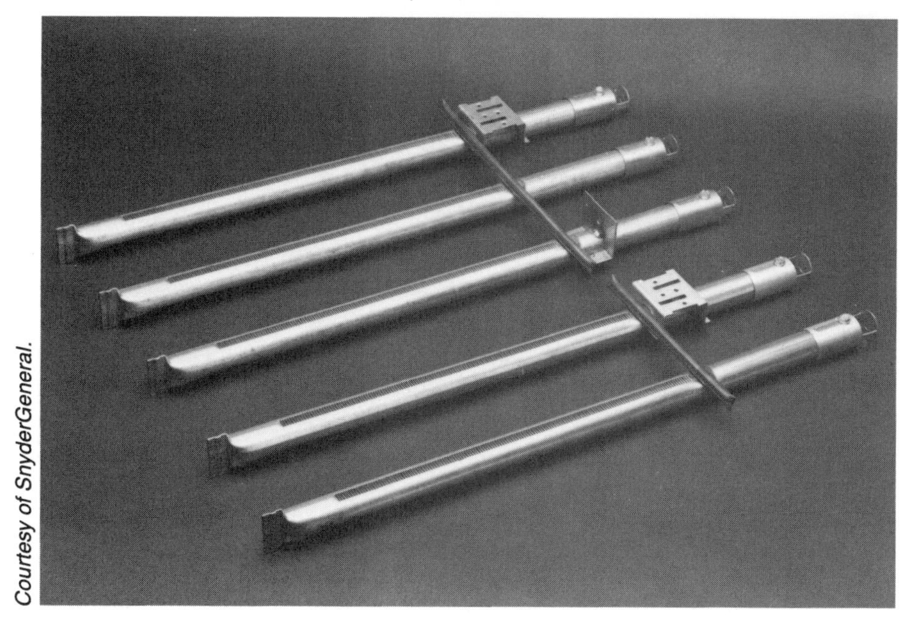
Courtesy of SnyderGeneral.

Fig. 5-3. *Ganged Gas Burners.*

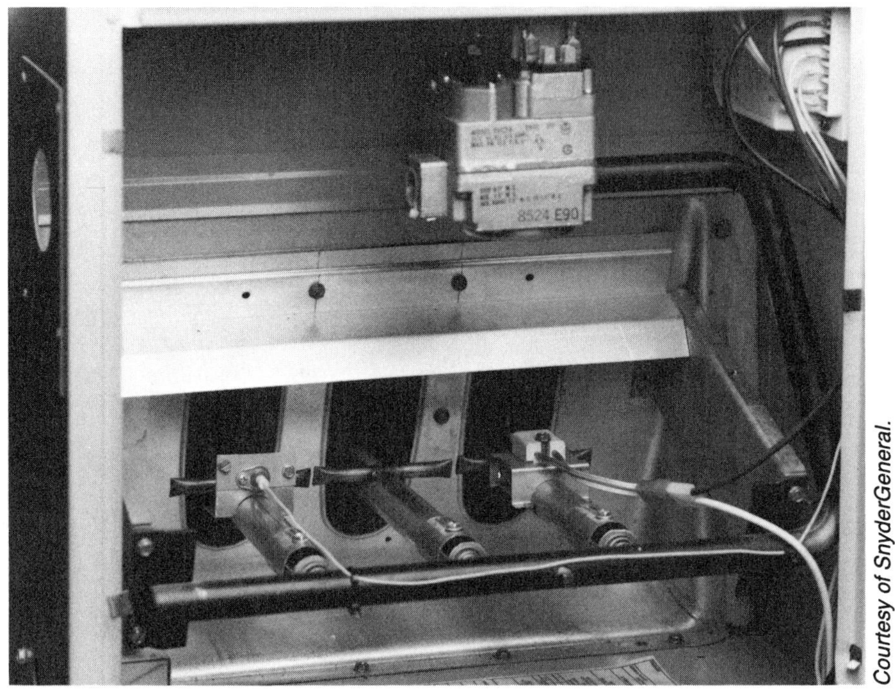

Courtesy of SnyderGeneral.

Fig. 5-4. A front view of a furnace showing the burners, manifold, and automatic gas valve.

HOT-WATER BOILERS

Hot-water boilers have a long and satisfactory performance record. Over the years, the system has been continuously improved. Many old systems circulated hot-water through cast-iron radiators by gravity. This method was neither efficient nor fast. Eventually, a more efficient circulating pump was added to the system which produced faster, more even heat throughout the building (see Figs. 5-5 and 5-6). Today, baseboard-type finned radiation eliminates the bulky cast-iron radiators and improves the appearance of the rooms (see Figs. 5-7 and 5-8). More modern systems include a hot-water boiler, with the necessary boiler trim and a circulating pump, plus all controls for the pump, boiler, and burner.

Hot-water heating systems can be completely installed by the homeowner without any ''professional help,'' but you will need additional tools: a pipe vise, pipe wrenches, cutting and threading tools for iron pipe, cutting and soldering equipment, solder, and soldering flux are needed for copper pipe. No sheet metal ductwork is needed. Copper pipe is commonly used because it is easy to work with. You will also want a pipe vise to hold copper pipe, although many home workshop vises have pipe-holding jaws and a pipe vise might not be needed. For a new installation, full installation instructions are furnished with the boiler, burner, and circulating pump. Because a cast-iron boiler is very heavy, the sections are shipped disassembled and must be joined using ''push nipples'' supplied with the package. Detailed instructions are furnished for this procedure. Follow these instructions carefully to eliminate leaks in the finished boiler. Small boilers might be shipped fully assembled, but you will probably need help to move the boiler to the basement.

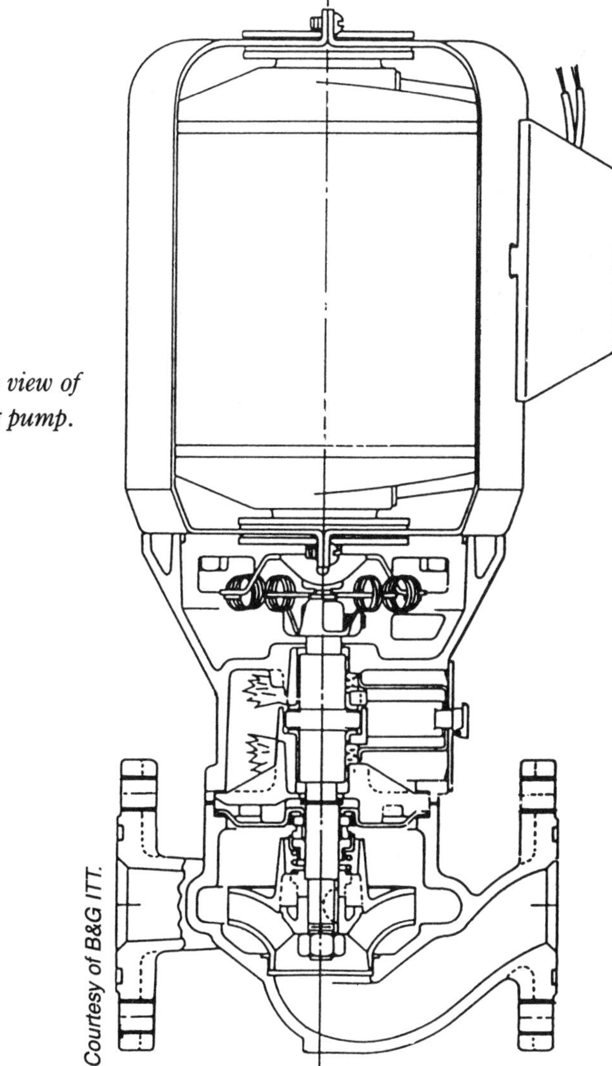

Fig. 5-5. *A cut-away view of a hot-water circulating pump.*

Courtesy of B&G ITT.

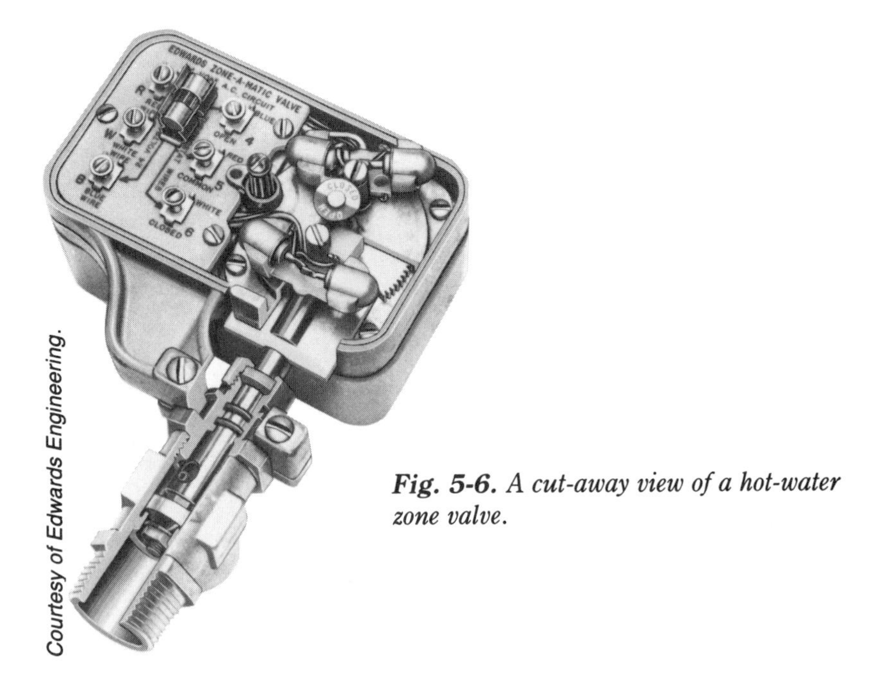

Courtesy of Edwards Engineering.

Fig. 5-6. *A cut-away view of a hot-water zone valve.*

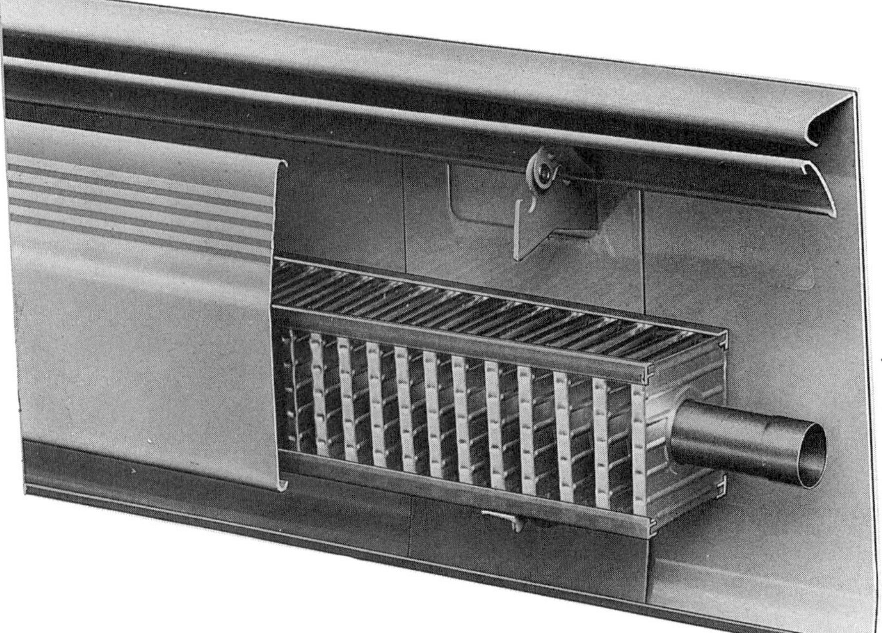

Courtesy of Edwards Engineering.

Fig. 5-7. *Finned, baseboard radiation for residential heating.*

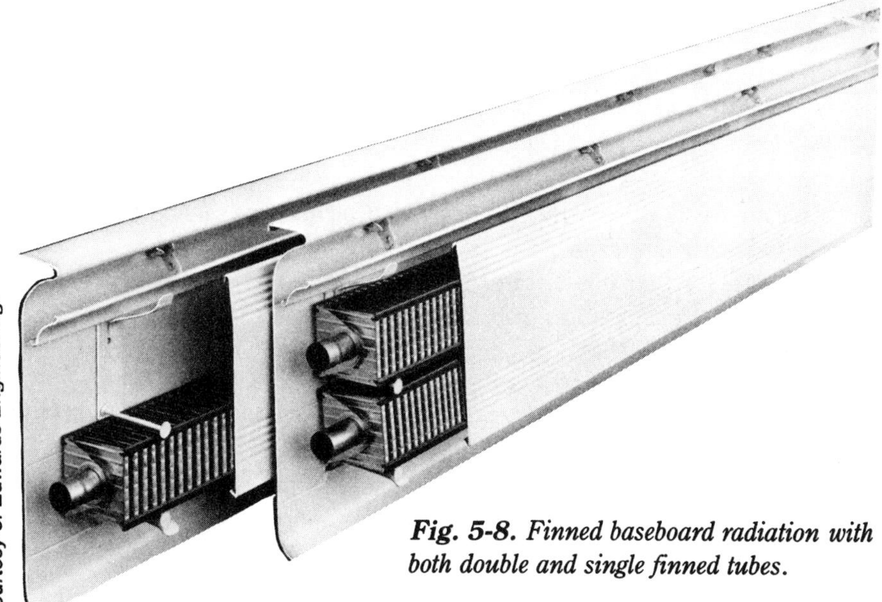

Fig. 5-8. Finned baseboard radiation with both double and single finned tubes.

Courtesy of Edwards Engineering.

If you are constructing a home and the finished walls are not in place, it is easy to run the piping to the baseboard radiation. If all of the walls are finished, however, it is more difficult, but perseverance will win out. If you must cut finished walls, cut openings very carefully so that patching will be minimized.

Although hot-water heating has many advantages, such as more silent operation, more even heat delivery, and no unsightly ductwork and furnace housing, the disadvantages are the difficulty and expense of installing central air conditioning as a separate system. If air conditioning is not installed, a separate, freestanding, portable humidifier must be used for winter humidification. Because of these disadvantages, you will want to think carefully before selecting this type of heating equipment. Be sure it meets all of your needs. For the average 1,600 square foot house, heating boilers are sometimes no larger than a good-sized suitcase. These boilers are very efficient, and combined with the hot-water circulating pump, make a very compact system.

RADIANT HEATING

In addition to hot-water radiant baseboard, there is radiant heating. This system consists of pipes carrying hot-water that are buried in concrete floors, or are concealed in plaster ceilings. While these systems are still being installed, their popularity has lessened somewhat. One reason is that this system requires that the thermostat be set about five degrees lower than baseboard radiation to achieve the same comfort. The most common complaint is that the floors are too warm. Other disadvantages include the possibility of leaks developing in pipes that are buried in the concrete or concealed in the ceiling material. Repairs are very difficult if the pipes are embedded in concrete floor. Systems can become air-locked and some areas might not heat satisfactorily. The elimination of air from these systems is difficult and pipes might need to be rerouted to ensure uniform heating.

Recently, two improved systems have been developed. One uses hot-water circulated through a half inch O.D. polybutylene thermoplastic tubing. The other system uses electric heating cable instead of tubing. Installation consists of covering tubing or electric cable with a poured gypsum mix called "Gyp-Crete 2000 Infloor Blend." This poured material covers and surrounds the tubing or cable that has been laid in a serpentine pattern, about four inches apart, on a subfloor. The electric cable is mounted on plastic strips, which hold the lines apart. The tubing is stapled to the underlayment. Only after the cable or tubing is installed and tested is the gypsum mix poured. The mix is leveled to 1¹/₄ inch thickness to produce a finished, level floor, ready for carpeting or vinyl to be applied. See Figs. 5-9 through 5-13.

Either type can be installed below grade if there is insulated board, followed by ¹/₄-inch plywood on top, then the tubing or cable is laid and covered with the Gyp-Crete, then the floor covering. **Note**: The concrete must have sufficient drainage to keep it dry.

Floor radiant heating systems do not require as high a thermostat setting as conventional systems. Usually, the setting can be reduced to 65°F or lower, and the occupants will still be comfortable. Floor temperature is usually about 85°F and the water temperature in the tubing ranges from 90°F to 140°F maximum. Water temperature can be adjusted at the zone control manifold.

These systems can be used for room additions or complete homes. The electric cable type needs no external heat source except electric power and one or more thermostats. The hot-water type requires an oil- or gas-fired boiler to supply the hot water.

Information on these systems and the location of local authorized contractors can be obtained from: Infloor Heating Systems, 920 Hamel Road, P.O. Box 253, Hamel, MN 55340, phone, (612) 478-6477.

HEAT SOURCES

So far, I have covered only the heat collecting and *delivery* components of the systems, the furnace or boiler. This section will cover the heat *producing* components: the burner, heating element (electric), and the heat pump. Heat *sources* used by the heat producing components are, starting with the lowest cost source, natural gas, liquefied petroleum gas (LPG), fuel oil, the heat pump, and electric resistance heating.

Which heat source you choose will depend on availability and cost. Many areas have both natural gas and fuel oil available. LPG (bottled gas) is usually available; its cost is between natural gas and fuel oil. The heat pump is efficient only in climates where the temperature in winter does not fall below 35° to 40°F. Electric resistance heating is the most expensive type of heat source available, costing three times as much as natural gas. In certain areas, the cost of electricity is subsidized by the federal government, and thus is much cheaper than the national average of 7.9 cents per kilowatt. Consequently, natural gas is the most economical fuel source, if it is available in your area.

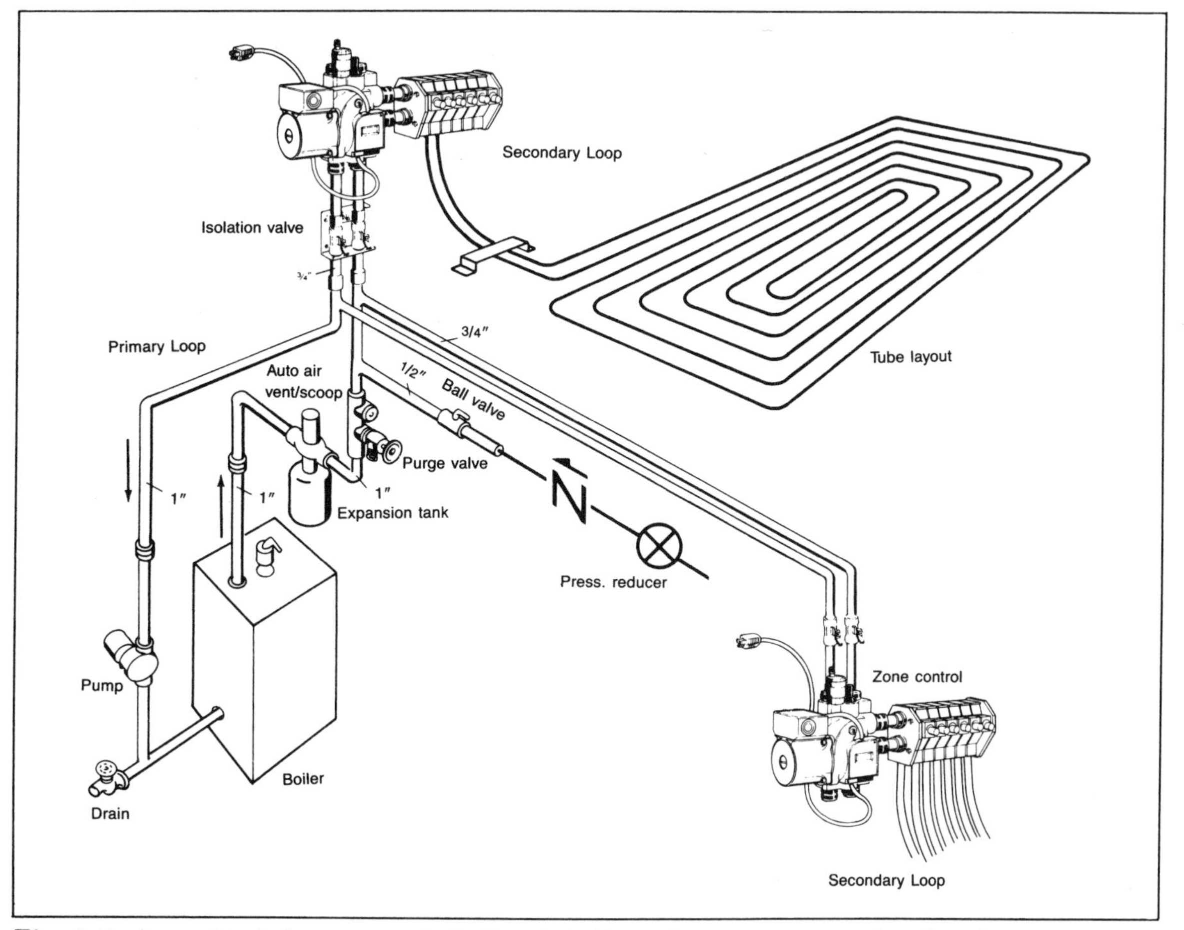

Fig. 5-9. *A complete infloor system, including the boiler and necessary accessories, for a hot-water system.*
Courtesy of Infloor Heating Systems.

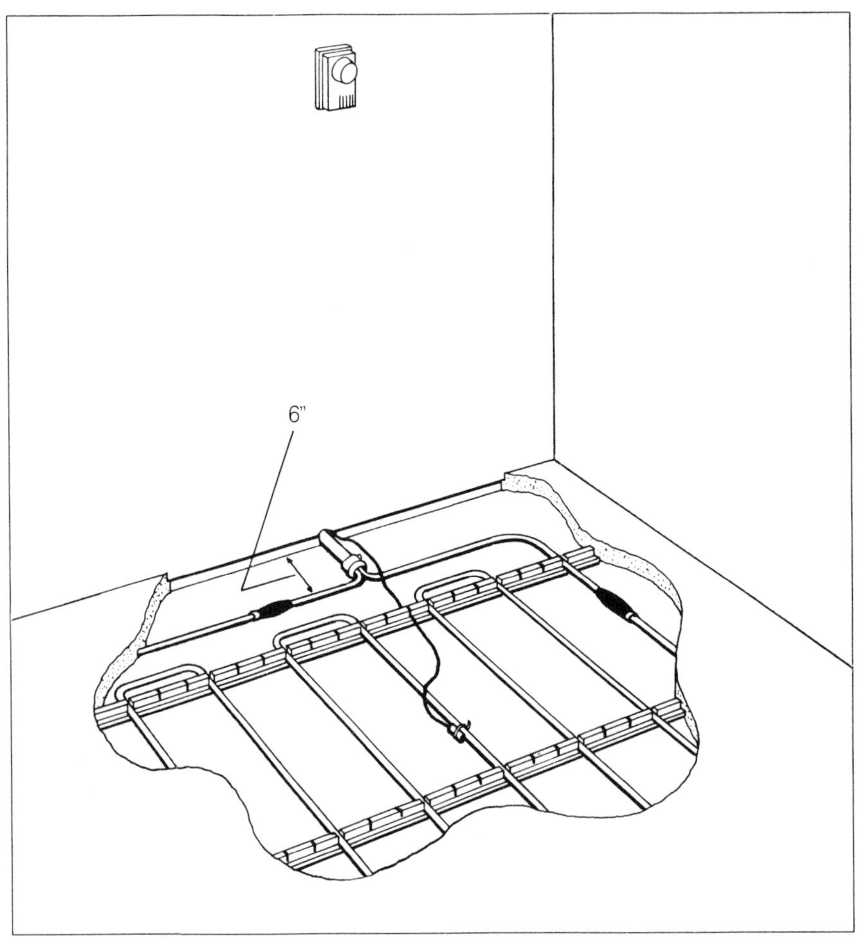

Fig. 5-10. The layout for electric cable. Courtesy of Infloor Heating Systems.

Natural Gas

Many well-established companies manufacture excellent heating and air conditioning products. Lennox Industries, Inc. in Dallas, Texas manufactures a very efficient "Pulse" forced, warm-air furnace that achieves an Annual Fuel Utilization Efficiency (AFUE) of up to 97 percent. Many furnaces are only 60 percent to 75 percent efficient, see Figs. 5-14 through 5-18. The Pulse furnace does not require a chimney, the flue gas is very cool, 130°F, and two-inch PVC pipe is used to exhaust the combustion products. This pipe can exhaust through a sidewall or through the roof. Condensate from the furnace can be drained to the city sewer, as it is not corrosive. In addition, there is no gas pilot that burns constantly. Instead, a spark plug ignites the gas. Lennox products are advertised widely and are readily available. Figure 5-19 shows an automatic gas valve.

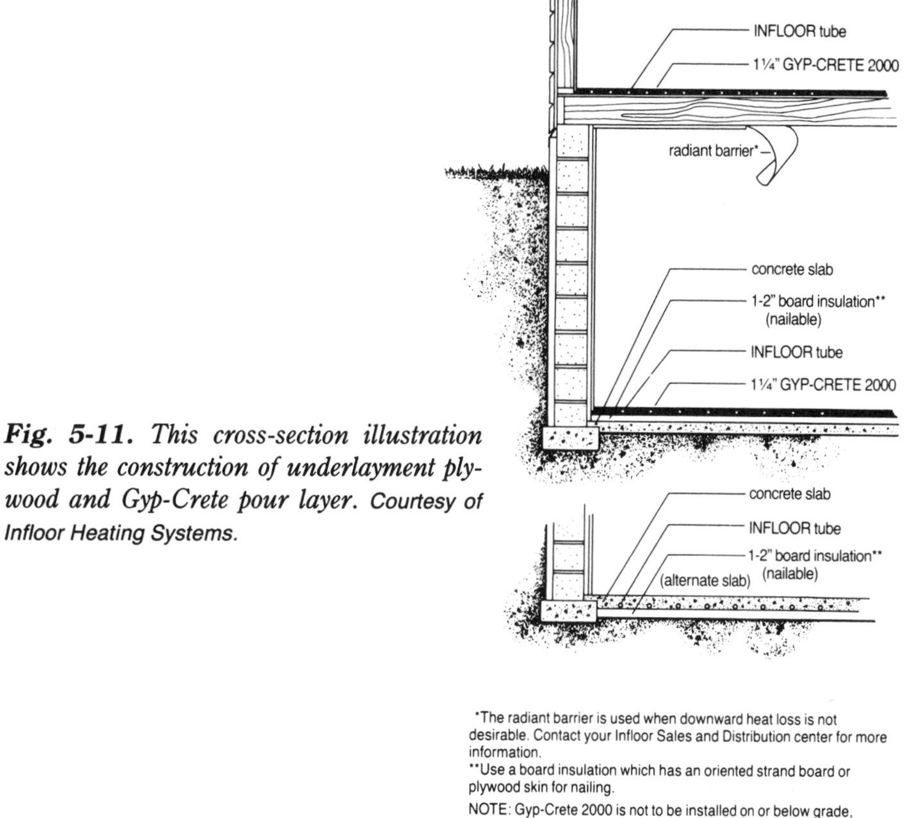

Fig. 5-11. *This cross-section illustration shows the construction of underlayment plywood and Gyp-Crete pour layer.* *Courtesy of Infloor Heating Systems.*

Lennox also manufactures excellent economical, standard gas-fired, forced warm-air furnaces and standard air conditioning systems. They also feature heat pumps. All items are readily available and are advertised widely.

Fuel Oil

Oil burners are comparable to gas burners except for the cost of fuel oil, which is about twice that of gas. A properly maintained oil-burning, forced warm-air furnace or boiler will operate as efficiently and cleanly as a gas-fired furnace or boiler. The oil burner requires more service, however, because there are more components. Annual service of the equipment is recommended. Even including the slight additional cost of maintenance, and the higher fuel cost, oil-fired equipment is the best choice when natural gas is not available. Reports of oil odors and oil residue on household items are the result of an incorrectly adjusted oil burner and are not common when proper maintenance procedures are followed.

Fig. 5-12. *Workmen spreading Gyp-Crete over heating tubes.* Courtesy of Infloor Heating Systems.

Fig. 5-13. *An infloor zone control and manifold with adjustment knobs and tubing leaving and returning to the manifold.* Courtesy of Infloor Heating Systems.

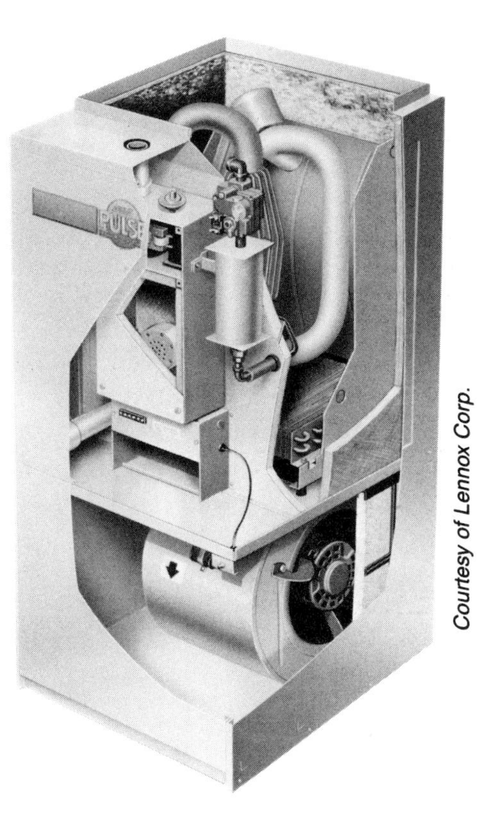

Fig. 5-14. *A Lennox Model G14 Pulse Furnace.*

Courtesy of Lennox Corp.

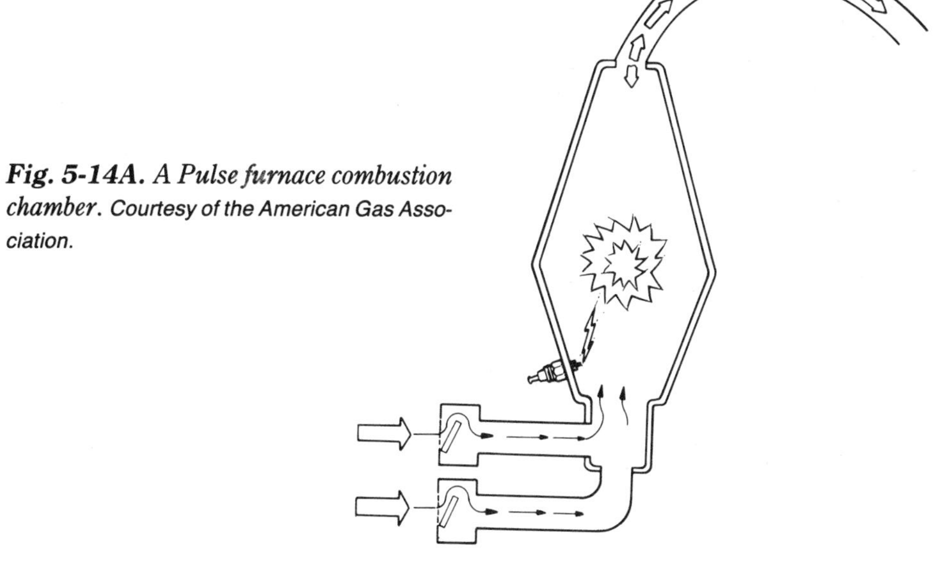

Fig. 5-14A. *A Pulse furnace combustion chamber.* Courtesy of the American Gas Association.

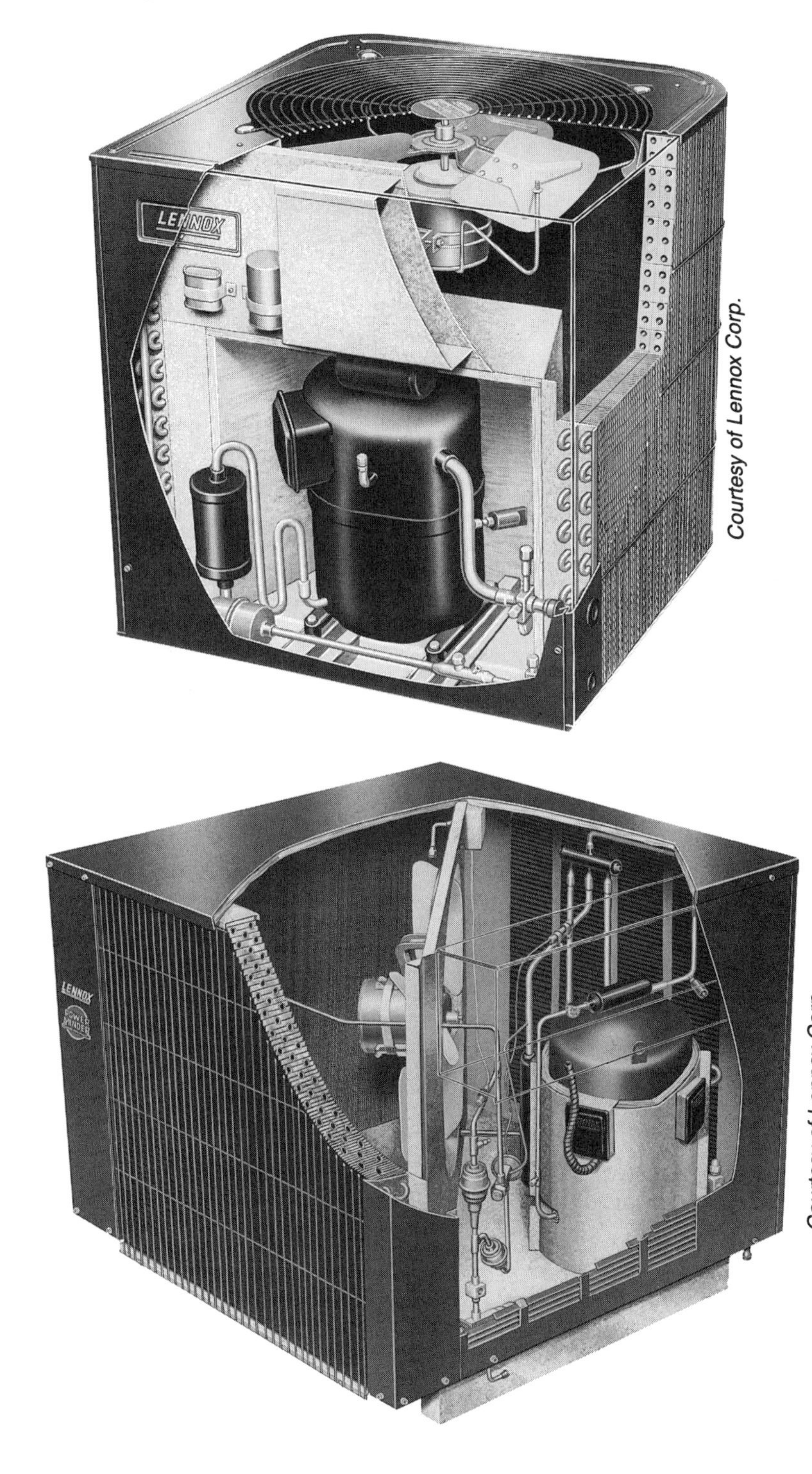

Fig. 5-15. A Lennox Model HS14 Two-Speed Condensing Unit.

Fig. 5-16. A Lennox Model HP14 Two-Speed Heat Pump.

Courtesy of Lennox Corp.

Fig. 5-17. *A Lennox Model HS19 Highest Efficiency Condensing Unit.*

The Heat Pump

The heat pump operates much like an air conditioning system, only in the reverse direction. The cooling coil in the furnace becomes a heating coil during cold weather. Similarly, the outside condensing coil becomes a cooling coil (see Figs. 5-20 and 5-21). The operating principle is that the cold, outside coil, being colder than the surrounding air, absorbs heat from the air. This heat is then "pumped" into the heating (condensing) coil in the furnace. As with the standard furnace, the blower (fan) blows this warm air throughout the house. The heat pump, by its nature, can only extract a certain amount of heat from the cold air. Therefore, as the outside air becomes colder, less heat is available to be "pumped" into the coil in the furnace. When temperatures drop to below 35°F, the heat pump becomes less and less efficient and supplemental heat is required in the form of electrical resistance heating coils or from a gas- or oil-fired warm air furnace. Electrical resistance heating costs three times the cost of gas heating. If a gas- or oil-fired warm-air furnace is already installed, then the heat

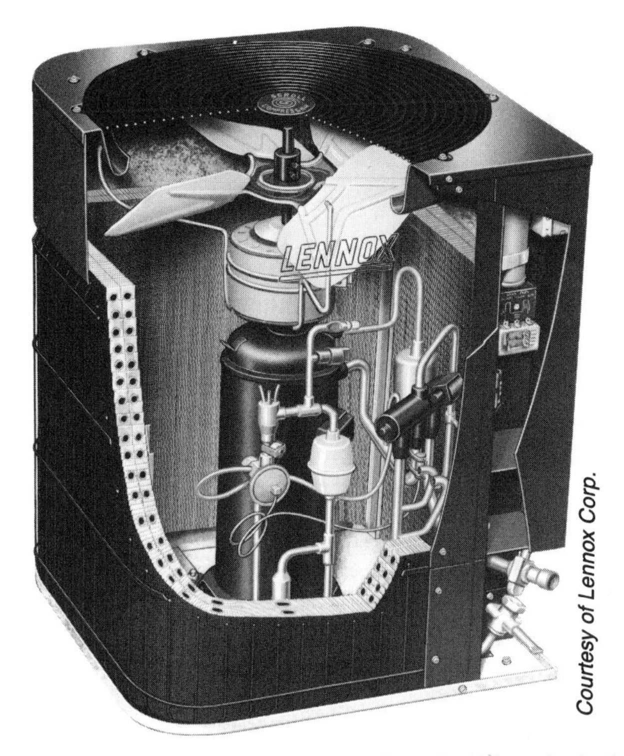

Fig. 5-18. *A Lennox Model HP20 Heat Pump with a "Scroll Compressor."*

pump should *not* have been purchased. Although the heat pump has been promoted by the electric utilities and heating and air conditioning contractors as an added plus, you don't need it. The heat pump is simply not suitable for all climates. While it is true that a heat pump system can be designed to operate in all climates, the expense of additional equipment to provide a sufficient source of available heat to the heat pump will outweigh any advantages gained.

Water-filled pipes, which are run underground to absorb ground heat, can be used as an additional source of heat by the heat pump. Unfortunately, this is very labor-intensive. Trenches about six feet deep must be dug for the piping system. The pipes are then fabricated and installed, for a total of about 300 feet, and finally, a circulating pump must be installed. The additional cost can easily equal the installed cost of the heat pump system. Other installations use a deep well for the heat source. Water from the deep well, once used, must be disposed of into a nearby dry well, and then you must consider environmental pollution. This is also very expensive and is not recommended.

The projected life of a heat pump is only eight to ten years, while an oil- or gas-forced warm-air furnace will last for 20 years. Don't be persuaded to purchase a heat pump system by a utility or contractor without first talking with a professional mechanical engineer.

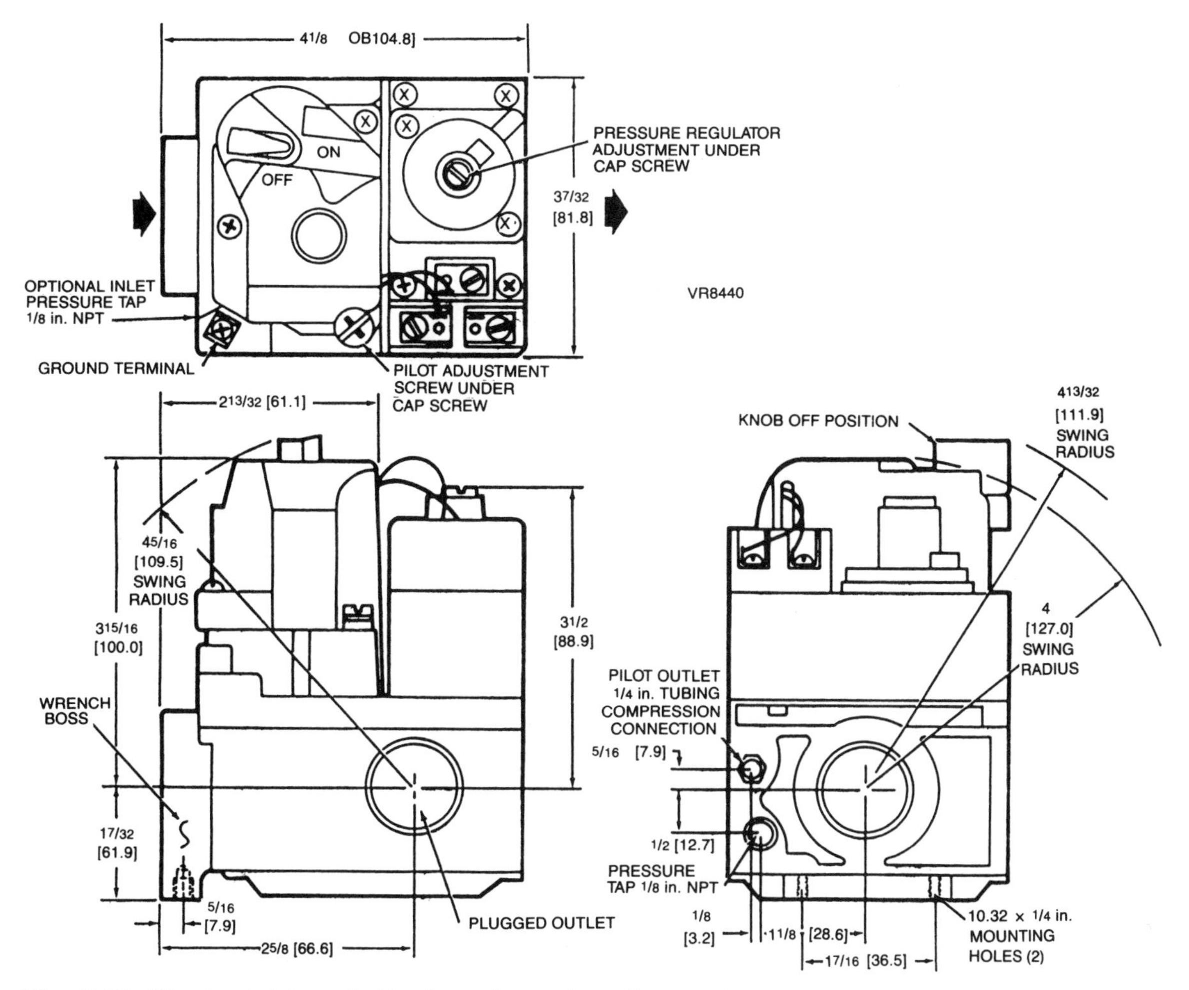

***Fig. 5-19.** The front, top, and side views of an automatic gas valve.* Courtesy of Honeywell Inc.

Electric Resistance Heating

The most expensive heating is electrical resistance heating, which is three times the cost of natural gas. Resistance heating consists of electric elements similar to those in a toaster or electric range. These coils are installed in ductwork and a blower (fan) blows air over them and into various rooms (see Fig. 5-22). Installation costs are lower than any other heating system, but operating costs are *extremely expensive*! As mentioned before, a few areas enjoy very low electric rates, but for the rest of the nation, rates average 7.9 cents per kilowatt, with the highest about 14 cents per kilowatt.

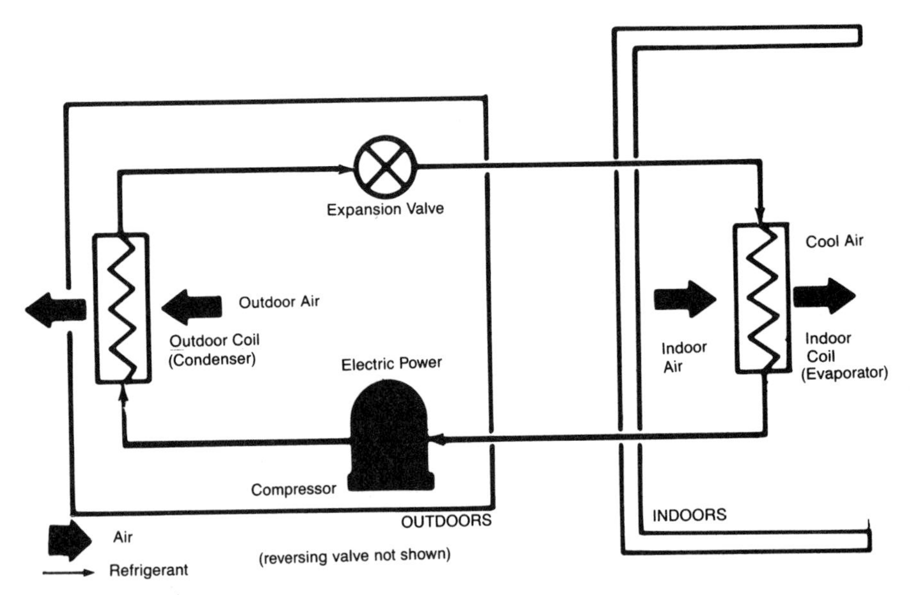

Fig. 5-20. The cooling mode for a typical air-to-air heat pump. Used by permission of the American Gas Association.

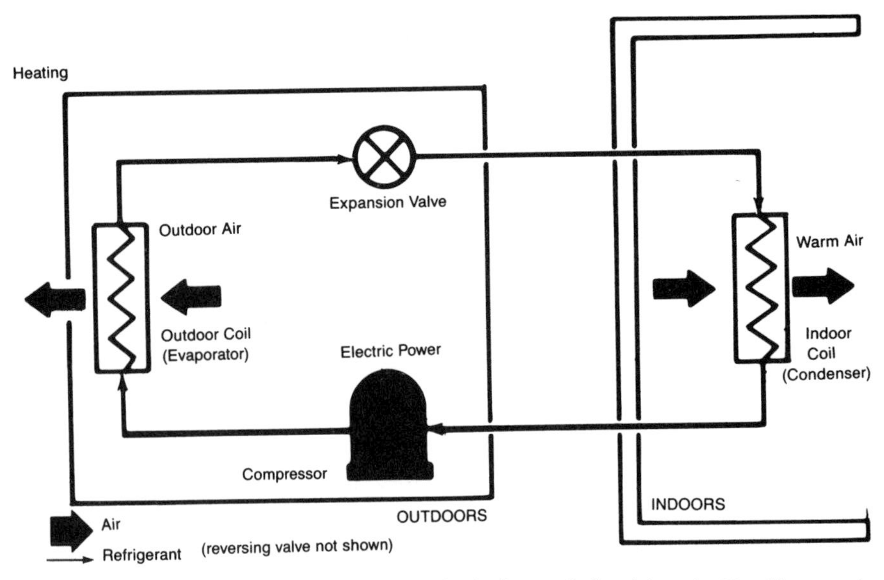

Fig. 5-21. The heating mode for a typical air-to-air heat pump. Used by permission of the American Gas Association.

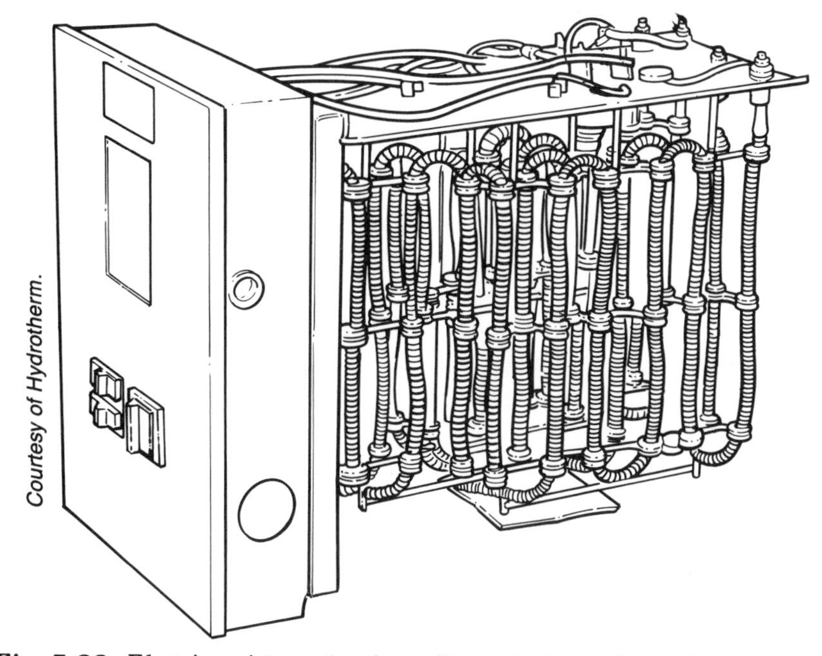

Fig. 5-22. *Electric resistance heating coils can be inserted into the duct work.*

The style of building has to be considered in addition to all of these other factors. A ranch style house built on a concrete slab will have the ductwork in the attic; these ducts must be insulated to prevent heat loss. Ducts already installed will have one or two inches of insulation. You would need to apply an additional two inches to reduce further heat loss. Later, insulation can be blown in to reduce heat loss through the ceiling. A total of 12 inches of insulation is recommended. Inspect the finished job to assure that a full six inches of insulation has been achieved. Houses with basements need extra insulation between the ends of the floor joists at the top of the basement walls. Small pieces of battinglike insulation, about six inches thick, pressed into these spaces around the perimeter of the basement can also help to reduce heat loss.

Another consideration is, if natural gas is in the area, is the gas line in the street in front of the property? If so, what is the cost of bringing a gas line into the house? These conditions are often found in partially developed subdivisions. The solution is to assess the cost to all properties along the new gas line route equally to pay for the extension of the gas line. Sometimes the gas, electric, and water lines are installed at no charge to the individual property owner, in this case the price of vacant lots includes the cost of the utility service.

With fuel oil, the situation is usually good if the house is in the city or surrounding suburbs. Fuel oil delivery companies are listed in the classified phone directory, and will usually accept new customers. Arrangements can be made for what is known as ''keepfull delivery.'' The oil supplier keeps the customer's tank full based on weather predictions. The oil tank is equipped with a whistle installed in the tank vent line. The whistle sounds as the tank is being filled, and stops when the tank is full. This alerts the delivery person to stop. Delivery can thus be made at any time and no one needs to be home. When credit is established, billing is monthly similar to a utility. This service is dependable; if an oversight does occur, the company will make a special delivery and restart the burner.

Making a $3,000 to $4,000 decision is no easy task. It is a good idea to select products from a reliable, long-standing manufacturer. Companies do change and some product lines do not compare with previous model years. Word of mouth from friends and acquaintances is a good recommendation. If you think you like a certain model of a certain make, ask the dealer to give you the names of current owners of that model. You can then determine if the customer is satisfied with the equipment and if the installation was satisfactory. Even though you select top quality equipment and contract for the installation or install it yourself, correct installation is as important as quality equipment. Go over the following checklist of elements you need to consider.

1) Capacity vs building heat loss or gain.
2) Compatibility of components. Is the capacity of the cooling coil matched to the capacity of the outside condensing unit?
3) Is the condensing unit located so that it will not be damaged? A shaded location is also best.
4) Is the ductwork sized properly? Are the hot-water lines sized properly and vented?
5) Have work permits been applied for (before any work is started)?
6) Is the contractor licensed and insured to do this type of work?
7) Has the finished installation been inspected and approved?

CENTRAL ELECTRICAL RESISTANCE HEATING

Electrical resistance heating is fairly reasonable to install. The *tremendous* expense is in the cost of operation. It is estimated that the cost of operating this type of heating, as compared to natural gas, is *three* times the cost of natural gas. The choice by all evaluations *is* natural gas. Electric heat helps the utility company even out its load throughout the year, but this does not give the electric customer any break. The residential space heating rate in the Detroit area is: Service charge per month, $6.15; all *on-peak* kwh at 11.26 cents each; all *off-peak* kwh at 6.26 each. This adds up to a monthly bill of $200.

Certain areas of the country have electricity furnished by government-operated generating plants that are powered by water. These electric rates are subsidized by the U. S. government and are very low compared to the average electric rate nationwide.

THE ELECTRIC FURNACE

The electric furnace is very clean, but *very expensive* to operate. In place of a flame, such as with oil or gas, electric resistance coils are built into the furnace or ductwork, see Fig. 5-22. These resistance coils are controlled either through a relay or directly if the thermostat is line voltage (120/240V). As a safeguard against the resistance coils burning out because no air passes over them because of a fan malfunction, a ''sail switch'' is installed near the coils. This switch has a 3″ × 3″ metal ''sail'' extending into the ductwork airstream. The sail is attached to a shaft; when the sail swings forward, contact is made causing the resistance coils to be energized. If the sail does not move, contact is not made and the coils cannot be energized.

If individual control of each room is desired, baseboard radiation resistance heating can be installed. Each room is then controlled by its own individual thermostat. This is more economical, as room thermostats can be turned down or off in unused rooms, thus saving on electricity. Figure 8-8 shows a line voltage thermostat for controlling electric heat.

Some through-the-wall air conditioning units incorporate a resistance heating coil and are able to provide some heat as needed. The thermostat is integral with the unit.

THE ELECTRIC BOILER

A boiler heated by electric resistance elements operates similar to the electric hot-water heater. Electric resistance coils insulated against the water are inserted through the boiler wall and made leaktight, either by a gasket and bolts, which hold it tightly against a mating flange welded to the boiler wall, or have external threads that screw into a tapped fitting welded to the boiler wall.

Built-in safeguards are exactly like gas- or oil-fired boilers: High-limit and operating-limit controls, a safety valve, a low-water cutoff water-feeder combination, and drain valve, for a steam boiler. For a hot-water boiler, controls are: Water regulating valve-relief valve, expansion tank, low-water cutoff water-feeder, drain valve, circulating pump, and pump relay.

All electric resistance heating boilers have the heating elements wired to be energized in sequence, one after another, at 60 second intervals. This is to prevent an overload on the electrical system. When the thermostat is satisfied, all the elements and the circulating pump are de-energized. The boiler wiring is also arranged so that if the high limit operates, all elements are de-energized and the circulating pump continues to run. This is to cool the boiler and elements down quickly.

Electric resistance boilers are manufactured in various sizes to meet the needs of different size buildings. Sizes range from 34,000 BTU/hour to 102,000 BTU/hour. The smaller models can be used to heat add-on rooms and larger models can heat an entire house. Other materials that are needed are: Sections of baseboard radiation to equal the length of all outside walls, available from home centers and heating dealers, who can calculate the lengths of radiation needed to equal the heat loss of the area that you plan to heat. Use copper tubing of a size to match the tubing in the radiation sections, fittings, solder, flux, and a torch to use for soldering the tubing and radiation and boiler into one system. A water supply must be brought to the boiler, a drain line must be provided to drain the boiler when necessary, also a line from the safety valve on the top of the boiler to a floor drain.

There are small portable and fixed radiator type units that are freestanding and require no installation except to be connected to an electrical circuit. These units are oil-filled and sealed and only require the setting of their self-contained thermostat.

If no other energy source is available, the larger size boiler is a good choice of heat. The 102,000 BTU/hour boiler easily heats a small to medium size house. Larger houses in milder climates could also be heated satisfactorily by this size boiler. Some boilers are completely packaged and wired, including the controls, leaving only the connections to water and power to make.

SELECTING THE BEST SYSTEM

Before you choose a system, evaluate it carefully. You'll want to get the most value and satisfaction for the investment. Some things to consider include: First cost. This depends on the building itself. Is it under construction or proposed; a completed building; rehabilitation of an old or historic building; and operating costs and maintenance. Hot-water heat and hot-air heat are widely different systems and require different installation techniques. For example, hot-water pipes can be easily concealed, similar to electrical wiring, usually in hollow partitions (sometimes called a chase, especially if formed for this purpose). On the other hand, large ductwork for forced, warm air cannot be easily concealed. This is especially true when working on old or historic buildings. Older buildings are constructed different than modern buildings, and installing ductwork can be difficult. Ducts can sometimes be hidden in closets or boxed into a corner. In most cases, hot-water heating pipes are able to be run through hollow partition walls to supply second floor heating equipment.

6

Installing an Oil-Fired Furnace

MODERN, OIL-FIRED FURNACES ARE MUCH MORE EFFICIENT COMPARED TO those made ten to fifteen years ago. Nonetheless, an oil-fired furnace is one of the more expensive heating units to install and operate. The price of fuel is about twice that of natural gas and the bulk of U.S. oil is imported, driving the cost up further. Another consideration of oil-burning systems is the availability of service.

In densely populated areas, oil delivery is very good. Sparsely populated areas and some rural areas usually do not have frequent oil delivery service, making it necessary to have a larger storage tank or two tanks. Most building codes allow two, 220 gallon oil tanks in a basement, provided that a "switching valve" is installed between the two tanks. This valve allows only one tank (220 gallons) to be connected to the oil burner at one time. This prevents a broken or leaking oil line from allowing 440 gallons of oil to flood the basement. If frequent oil delivery service is available, one 220 gallon tank is sufficient for a 1600 square foot house.

REPLACING THE OIL BURNER

There are situations where the complete furnace or boiler is in excellent condition, perhaps it was recently replaced. Consequently, the oil burner is probably much older than the furnace or boiler, and is now inefficient and obsolete. In this case, it is best to contact the furnace or boiler manufacturer or a local dealer representative and order a burner designed for the furnace or boiler that you have.

Replacing only the burner can save you a lot of money. You won't have to change the ductwork of a warm-air furnace or the piping of a boiler. If your system is ten to fifteen years old, however, consider replacing the complete heating system. If you decide to install a complete furnace replacement, this is the best time to install central air conditioning. Ductwork and piping for the connection of the cooling coil to the condensing unit can be properly done. New furnaces come with the proper fan for air conditioning already installed at the factory.

Modern, very high-efficiency oil and gas furnaces, up to 96 percent efficiency, are much more expensive than the 80 percent efficiency standard type. The payback is longer, possibly ten years, before any savings will be realized. You must decide between the various types and efficiencies available. High-efficiency furnaces are more complicated and require more maintenance and monitoring than standard heating equipment. Various types of high-efficiency oil and gas heating equipment are shown in this chapter. A high-efficiency oil furnace is shown in Figs. 6-1, 6-2, and 6-3. Figure 6-4 shows various fuel oil pumps.

Modern oil burners are much improved over previous styles. The blast tube has been redesigned by the addition of a flame retention section. Certain localized conditions can cause the flame to leave the burner head and "float" in the combustion chamber. When this happens, the flame is often extinguished. If the flame does not relight, the safety control locks out the burner. Most burners now have constant ignition to maintain a flame and to reignite the burner upon flame failure. Constant ignition provides a constant spark to reignite after flame failure. These improvements make oil heat a very clean, satisfactory home heating system.

It is strongly recommended that an "oil fire valve" be installed in the oil supply line at the burner. The oil fire valve is a special valve designed to snap shut in case of fire at the burner. It senses an unusual temperature increase at the burner location.

INSTALLING A NEW BURNER AND FURNACE

If you will be installing a complete new burner and new furnace, you need to remove the old equipment and clean the area to make ready for the new equipment. You will also need to do a few other things.

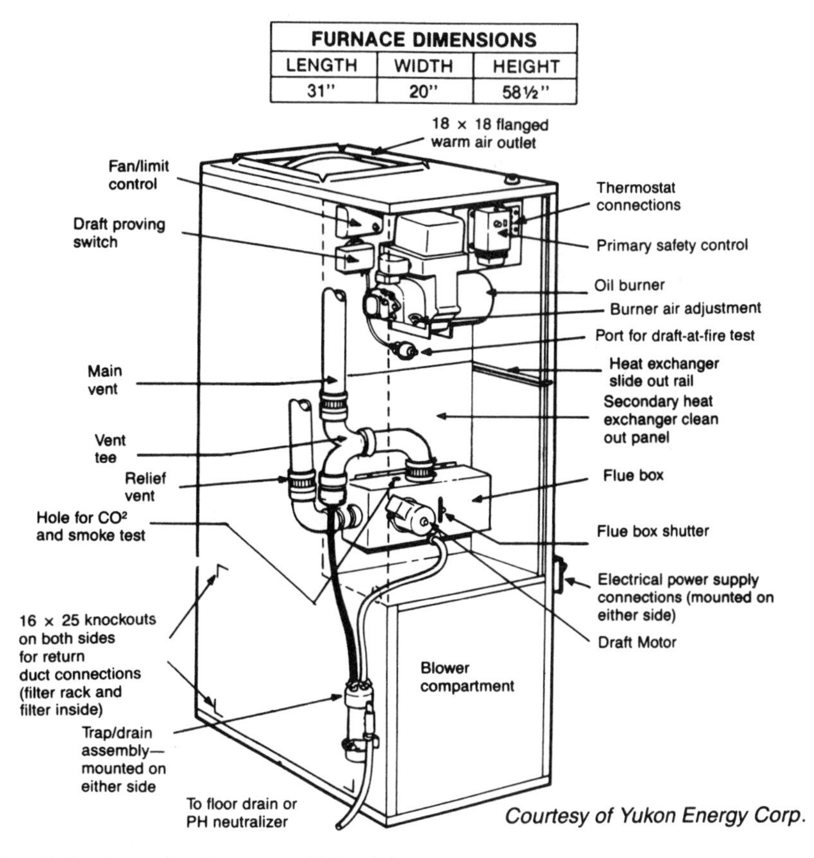

Courtesy of Yukon Energy Corp.

Fig. 6-1. *A condensing type oil-fired furnace.*

Adapt the Ductwork

You need to adapt the present ductwork connections to fit the new furnace discharge and return duct openings. It is best to contact a sheet metal fabrication shop to make these adapter pieces. Have the shop send someone to measure for the necessary pieces. Generally, the furnace discharge goes straight up to meet the trunkline (the lengthwise duct that goes across the basement ceiling). A flexible connection of flame-proof material is installed between the furnace and the ductwork to prevent vibrations from being transmitted to the ductwork.

You might have to redirect the return air ductwork if the return opening on the furnace is in a different location. Some new furnaces have panels stamped so that the opening can be on one of the three sides. This might eliminate redirecting the return air duct, except for the final duct-to-furnace connection.

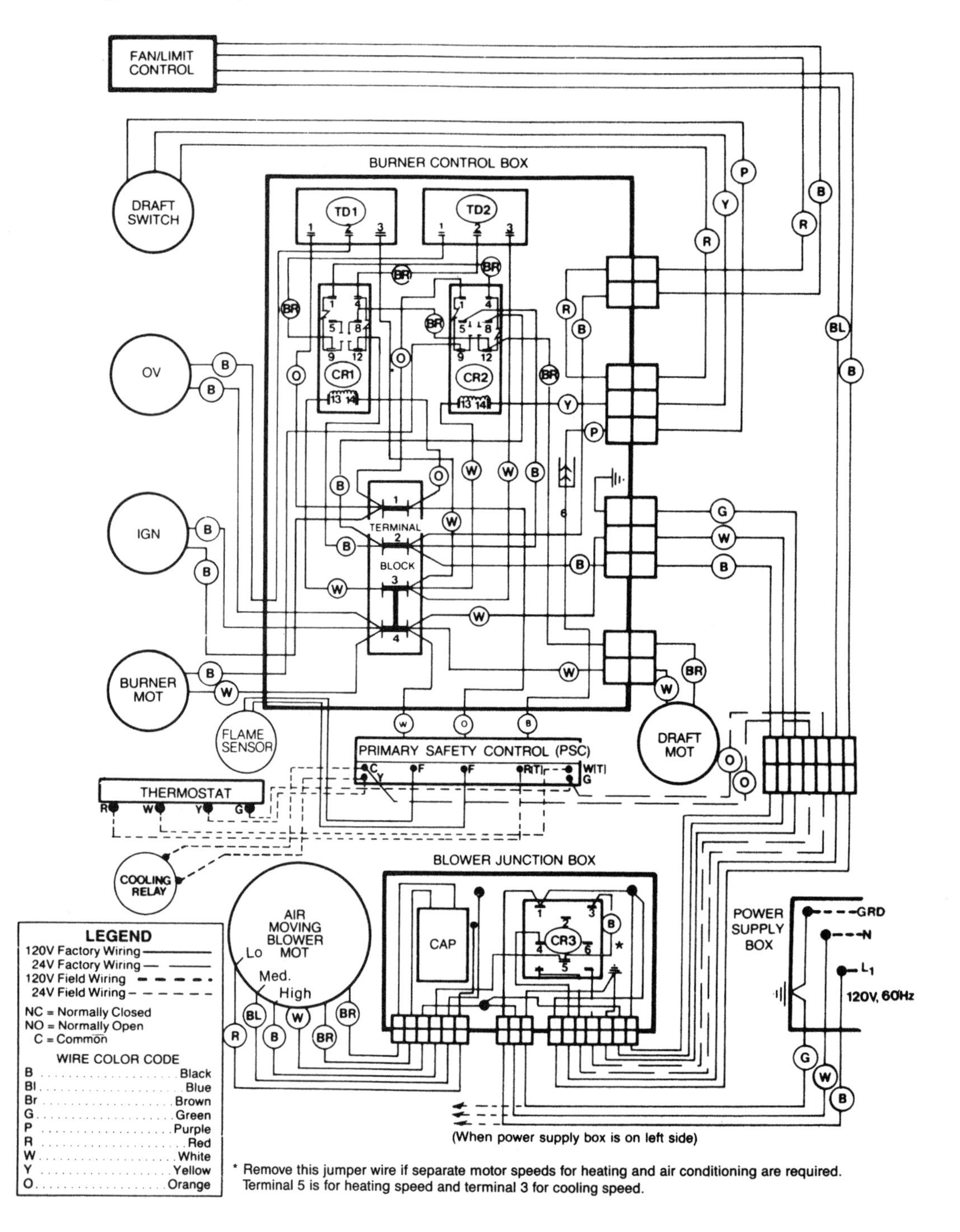

Fig. 6-2. A wiring diagram of the furnace shown in Fig. 6-1. Courtesy of Yukon Energy Corp.

HEATING

1. On a call for heat, the oil burner primary safety control (PSC) energizes, if the high temperature limit and draft switches are closed (normal).

2. The PSC energizes the control relay (CR1), the pre-purge timer (TD1), the ignition transformer (IGN) and the oil burner motor.

3. CR1 in turn energizes the draft motor.

4. After 15 seconds TD1 energizes the oil valve (OV), at which time the burner should light.

5. The burner will operate until the heat demand is satisfied, at which time the PSC de-energizes CR1, TD1 (and oil valve), IGN, and burner motor.

6. At burner shutoff the draft motor will continue to be powered through the post-purge timer (TD2), which is initiated through the normally closed contact of CR1. TD2 powers the draft motor for 5 minutes after the burner shuts off.

7. As the furnace warms up, the fan control will power the air moving blower motor, and shuts it off as the furnace cools after the burner has been shut off.

8. If the limit control functions to open the burner circuit because of high temperature the oil burner is shut off but the draft motor continues to purge the flues, being powered through TD2.

9. If the draft switch functions, it opens the oil burner circuit to shut off the oil burner, and energizes control relay (CR2) which energizes both the burner motor and the draft motor, which will continue to run indefinitely. If the loss of draft is temporary, the burner circuit will be remade and there will be an attempt to relight if the thermostat is still calling for heat.

COOLING

1. On a demand for cooling, the cooling relay and fan relay CR3 are energized through terminals Y and G of the PSC.

2. CR3 energizes the circulating air blower motor speed selected for air conditioning and disconnects the heating speed, if different.

SCHEMATIC AND CONNECTION WIRING DIAGRAM

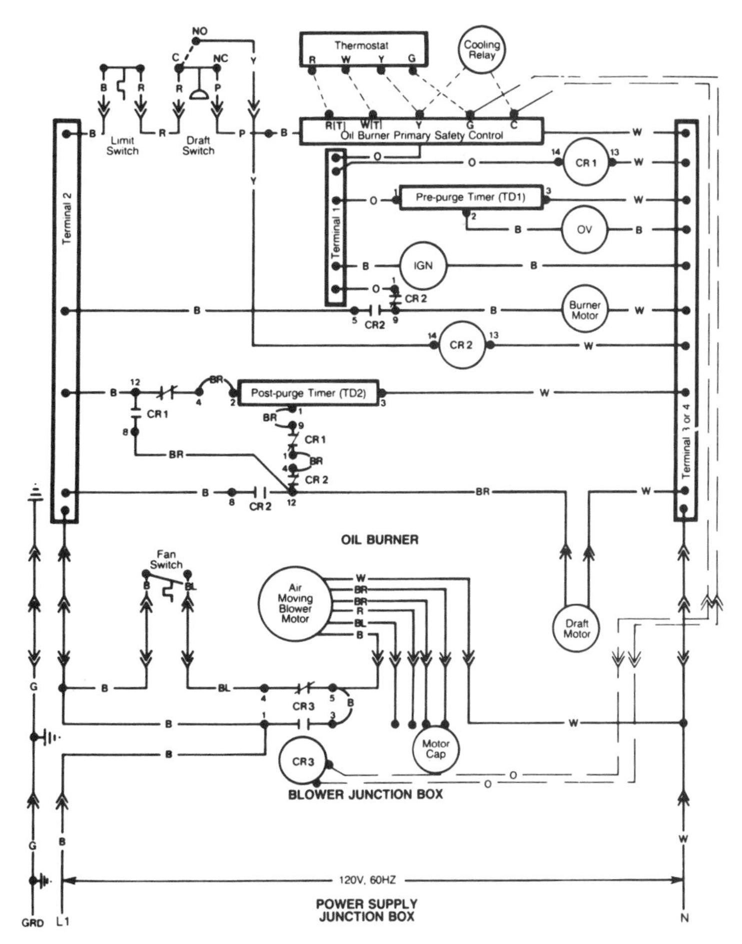

Fig. 6-3. A schematic wiring diagram, including the sequence of operation. Courtesy of Yukon Energy Corp.

Fig. 6-4. Different styles of fuel oil pumps, each designed for a specific use. Notice the information printed on the nameplate. Courtesy of Suntec Company.

Adapting the Oil Burner

The conversion type oil burner can be adapted to most forced, warm-air furnaces. Many areas do not have natural gas available and the oil burner is the next best heat source. The operating cost is midway between natural gas and electric resistance heat.

An oil burner requires a combustion chamber to contain the flame. This chamber is fourteen inches to eighteen inches in diameter and can be made of stainless steel or insulating firebrick. The stainless steel chamber is sized to match the firing rate in gallons per hour (GPH). Some firebrick chambers are preformed to size, while others are built on the job to fit the available space. These preformed chambers are used when they can be installed in the firebox. Oil burner suppliers will have combustion chambers and other accessories.

Information needed to purchase an oil burner can be found on the furnace nameplate. The nameplate gives the BTU input and output, such as, 85,000 BTU, 100,000 BTU input. The BTU input determines the firing rate in gallons per hour (GPH). The standard domestic oil burner uses No. 2 fuel oil, which contains 136,000 BTU per gallon heat value. Most single homes of up to 1,800 square feet floor area would have a 150,000 BTU furnace fired at a rate of 1.10 GPH. Smaller homes might need a firing rate of 0.70- or 0.85-GPH.

If you are installing a new oil burner, inspect the combustion chamber for replacement, if needed. Be sure to measure the present chamber for correct size replacement. The furnace manufacturer's local dealer should have all of the necessary parts on hand, or he can order them for you. In addition, you need furnace cement, bolts, 3/8-inch copper tubing, 3/8-inch flare fittings and flare nuts, a short piece of flexible metal conduit, necessary fittings, and wire connectors. A new stack safety relay or other safety control might be included with the new burner. The fan and limit controls can still be used, if in good operating condition.

Determine that the wiring to the furnace is on its own separate circuit. If the wiring on the furnace is in good condition, it can be reused. Replace the flue pipe and vacuum, and brush the passages inside the furnace. Determine the nozzle firing rate and angle of spray, and whether it is a solid or hollow spray. A designation will be stamped on the nozzle such as, 0.85-80-H, meaning 0.85 gallons per hour oil flow, 80° spray angle, and H, hollow or S, solid. The hollow spray is similar to a funnel, with the funnel wall the oil droplets. A solid spray is still the cone shape, but the cone is solid with oil droplets.

The design of the replacement burner might be slightly changed, making minor adjustments necessary. Other than this, there should be no problems installing the burner in the furnace.

Setting the Burner

Determine how the burner is attached to, or is fitted into, the combustion chamber opening. An exact replacement is usually flange mounted to the combustion chamber opening. A replacement or conversion burner can be pedestal mounted. The pedestal must be anchored to the floor with concrete anchors. Before starting the installation, open the rear of the blast tube and remove the nozzle holder/oil tube assembly and electrodes, and set them aside. This prevents damage to these parts while you install the blast tube and burner in place.

Wiring the Burner

Once you have the burner body in place and all of the parts assembled, the wiring to the burner can be done. With most burners, the wiring goes first to the stack relay, then to the burner body. If an electric eye is used instead of a stack relay, the relay for the electric eye is usually mounted on the burner body.

If the safety relay has been in use for many years, it is probably wise to buy a new replacement. Take the old relay with you when you buy a new relay. In most cases, the old relay can be traded in for a new one, and you can save 10 to 20 percent on the full price. Installation instructions and start-up methods are furnished with all new burners.

The burner wiring connections are generally in one end of the ignition transformer case. There are two knockouts and a small cover for the wiring compartment. One knockout is for the wires from the stack relay, the other is for the wires going to the oil burner motor. Two or three wires will come from the stack relay: one white, the neutral; one black, the hot wire (this wire will feed the oil burner motor *and* ignition transformer if constant ignition is specified); a third wire, usually red, will feed the ignition transformer only, if intermittent ignition is specified. If a new relay is furnished with the replacement burner, the type of ignition action will be built into the relay.

Constant ignition provides a spark as long as the burner is in operation. Intermittent ignition provides a spark for a set length of time. If a flame is established and maintained for this set time, ignition is stopped. Some burner installations do not maintain a proper flame because of changes in chimney conditions or other reasons. Constant ignition can be easily provided. Disconnect the ignition lead from the IGN terminal at the relay and tape the bare end. Do the same at the end in the transformer compartment. Connect the transformer lead and motor lead to the black wire from the relay (this wire is on the MOT terminal of the relay).

Before reinstalling the ignition/nozzle holder assembly, reach in and turn the blower fan wheel. It should turn, but with some resistance, as the fuel pump is attached to the same shaft. After installing the ignition/nozzle holder assembly, connect the ignition electrodes to the ignition transformer terminals. Many types of connections are used: insulated wire, similar to spark plug wire with snap-on ends; and stiff wires on the ignition electrodes that make contact with small, flexible plates on the transformer terminals. The transformer opens on hinges and, when closed, presses on the stiff wires with its flexible contact plates. Connect the oil lines to the burner as required by the instructions. Caulk any openings around the blast tube with an approved furnace cement. Refer to Figs. 6-5 and 6-6.

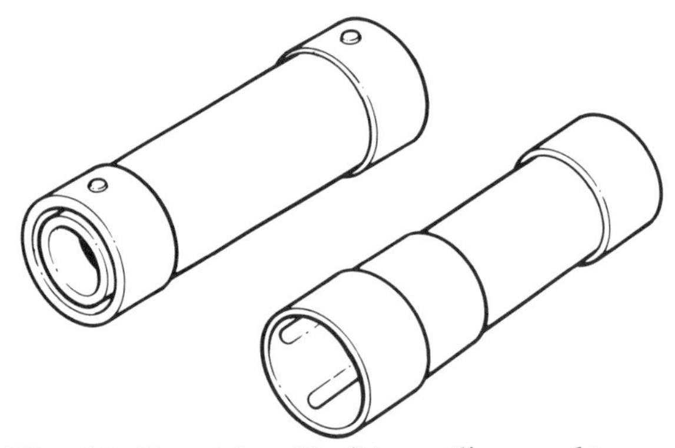

Fig. 6-5. *Two styles of flexible couplings used to connect the motor to the fan and the oil pump. The centers are made of a flexible rubber.* *Adapted from a catalog illustration—Westwood Products Co.*

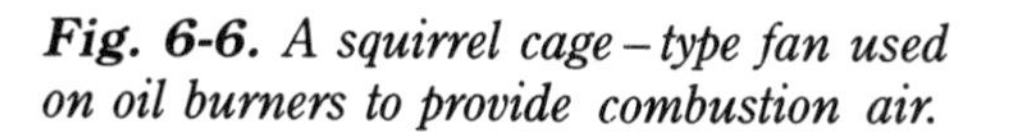

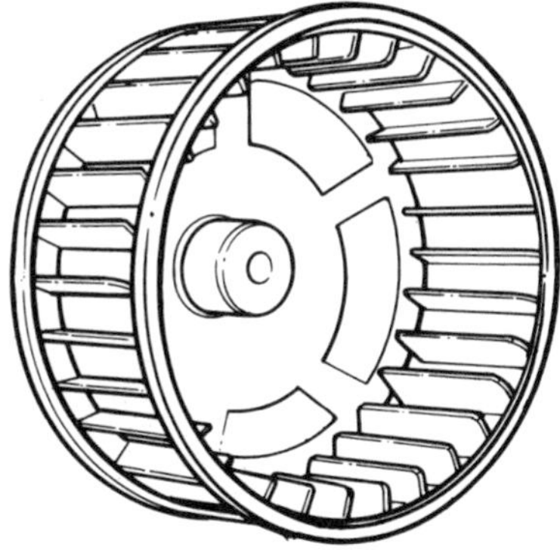

Fig. 6-6. *A squirrel cage–type fan used on oil burners to provide combustion air.*

If you are not familiar with oil burner operations and start-up, it would probably be wise to have the burner started and adjusted by a service technician. Necessary tests must be made, for which specialized instruments are needed, and it is worth the cost of the service call to have the start-up done properly.

Furnace controls used on the furnace are limit controls, set to shut the burner off at 200°F, a fan control, set to start the fan at 130°F and turn it off at 90°F. Many times, these controls are combined in a single case and share the same bi-metal coil that actuates both controls.

7

Installing a Replacement Boiler

OLDER BOILERS GAVE EXCELLENT SERVICE IN THEIR TIME. ALTHOUGH NOT AS efficient as today's boilers, they operated with little maintenance. These cast-iron boilers usually had a conversion gas or oil burner installed in them. Their low efficiency of about 60 percent, however, make them excellent candidates for replacement. Modern gas-designed or oil-designed boilers can easily achieve an efficiency rate of 80 percent.

Installing your own new boiler-burner combination could possibly pay back its cost in about three years. Even with the expense of some professional help, a four year payback is still possible. The hardest part of the replacement job is disconnecting and removing the old boiler. Follow the directions and illustrations in this chapter, and the job will be easier. Figures 7-1 through 7-5 illustrate the modern hot-water boiler, showing the flues, and two views of insert hot-water (domestic) coils exposed.

REMOVING THE OLD BOILER

Connecting a new boiler to current hot-water lines or steam lines can be easily handled by the handyperson. Forced, warm-air systems entail sheet metal work that the handyperson is not equipped to handle nor fabricate. See Figs. 7-1, 7-2, and 7-3.

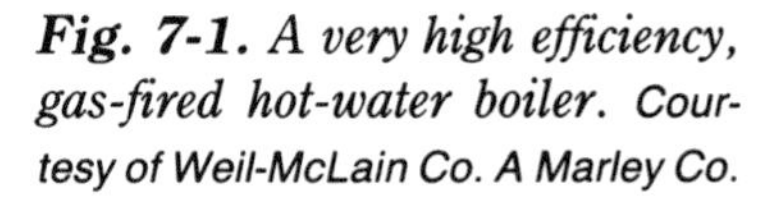
Fig. 7-1. A very high efficiency, gas-fired hot-water boiler. *Courtesy of Weil-McLain Co. A Marley Co.*

Tools needed to install a new replacement boiler and remove an old one can usually be found in the home workshop. You will probably want to rent two, two-foot pipe wrenches, a pipe cutter, a pipe reamer, pipe dies, and a pipe vise. This remodeling work is hard work; proceed carefully and do not injure yourself. Wear safety goggles when breaking up materials or doing any work above your head, or where particles could fly into your eyes. I speak from experience; wear safety goggles. Figure 7-8 shows a boiler section.

Removing the Riser

The pipe leaving the top of the boiler, called a ''riser,'' is usually a two- or two and a half-inch pipe. After breaking the el, unscrew the riser counterclockwise. Fit the two-foot pipe wrench on the riser near the boiler in a position so that you are able to pull on the handle (toward you). If the riser does not budge, find a pipe that will fit over the wrench handle. This ''cheater'' will give you extra leverage and should be two or two and a half times the length of the wrench handle. Do not allow the cheater to slip off the wrench handle. Another remedy is to use penetrating oil such as Liquid Wrench, that is available at hard-

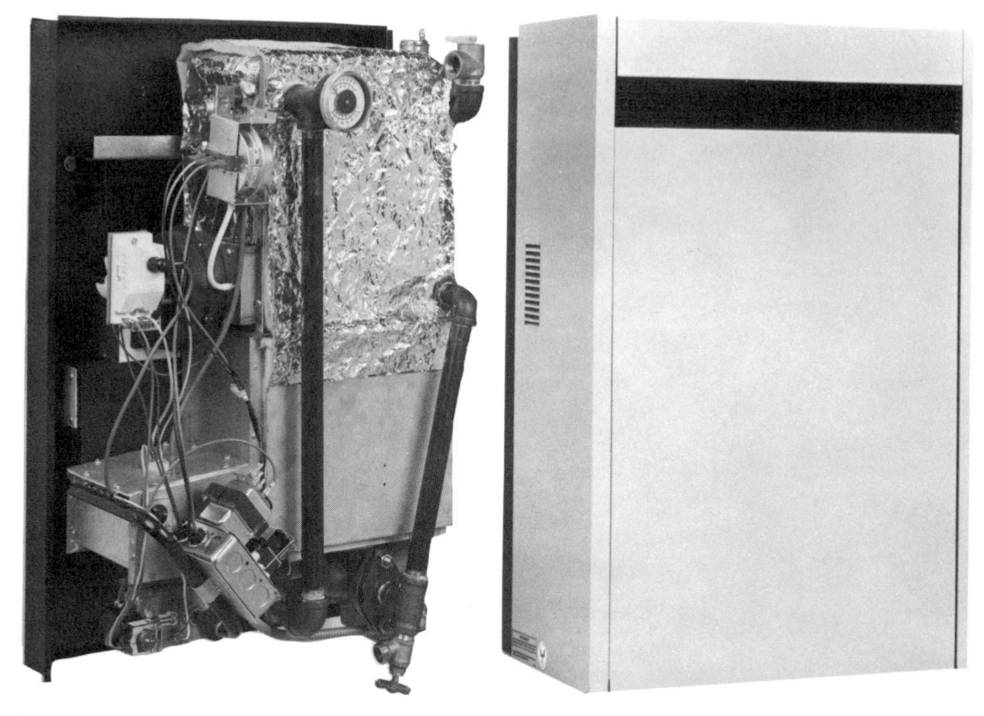

Fig. 7-2. *A cut-away view of a fully self-contained, gas-fired, hot-water boiler. Courtesy of Weil-McLain Co. A Marley Co.*

ware stores. Squirt the oil on the threads just above the boiler connection and allow to set overnight. Try again the next day. Caution: When pulling or pushing down on any tool, protect your knuckles by pushing on the handle with the heel of your hand, palm open. Lightweight ''bump hats'' are available at home centers to wear in places having low or sharp objects in the way. This is *not* the ''hard hat'' used on construction projects.

Disconnecting the Water Supply

After removing the riser from the boiler, disconnect the water supply. This line will be either 3/4 inch or 1/2 inch. A correct installation has a union between the shutoff valve and the boiler. If there is no valve, shut the water off at the meter and cut the pipe with a hacksaw. You will need to buy a shutoff valve to fit the pipe, plus two, three-inch nipples. Cut the pipe close to a fitting and remove the short piece. Install one of the nipples in the valve using thread compound or Teflon tape, and tighten in the fitting snugly. It is wise to install a pipe plug in the open end of the valve until you're ready to connect the line to the new boiler. Boilers are shown in Figs. 7-9 and 7-10.

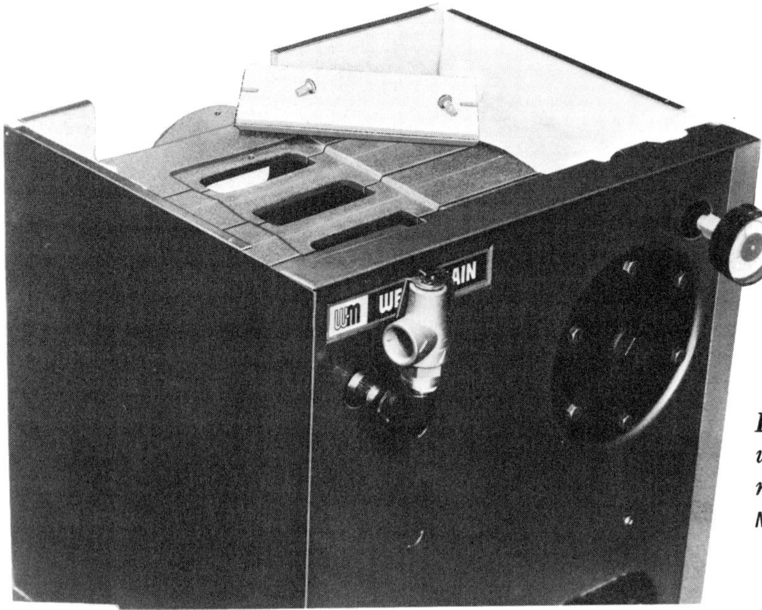

Fig. 7-3. *To clean the flues of a hot-water boiler such as this one, just remove the cover.* Courtesy of Weil-McLain Co. A Marley Co.

Fig. 7-4. *A hot-water boiler showing the insert coil that provides hot water year round.* Courtesy of Weil-McLain Co. A Marley Co.

Fig. 7-5. *A large, hot-water boiler with a large, domestic hot-water insert coil for increased hot water use.* Courtesy of Weil-McLain Co. A Marley Co.

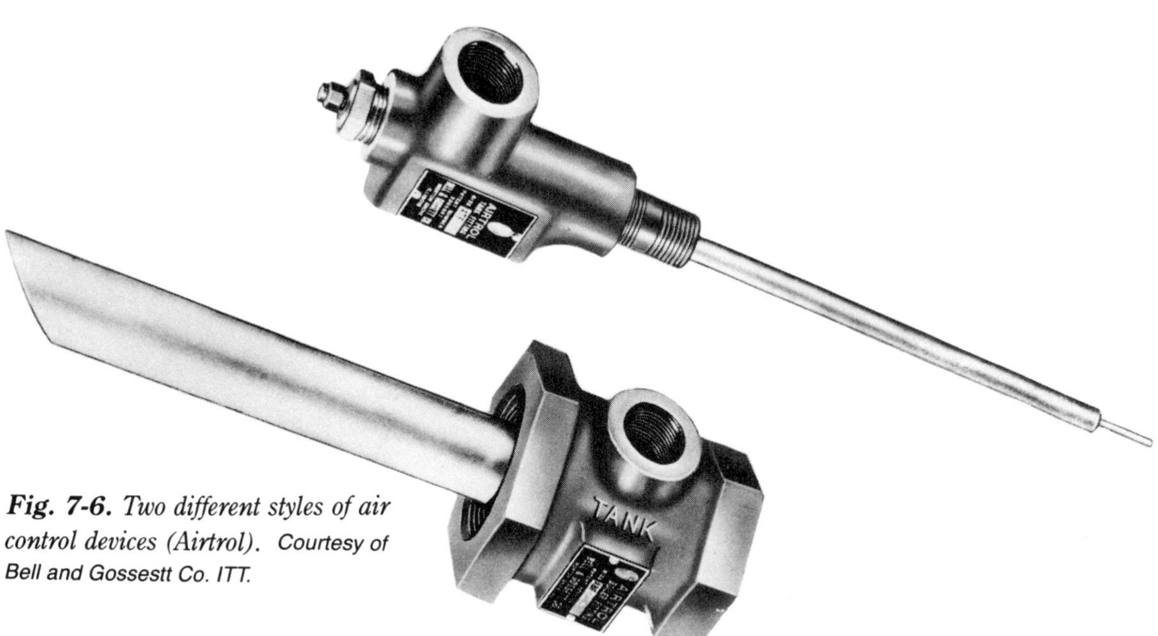

Fig. 7-6. *Two different styles of air control devices (Airtrol).* Courtesy of Bell and Gossestt Co. ITT.

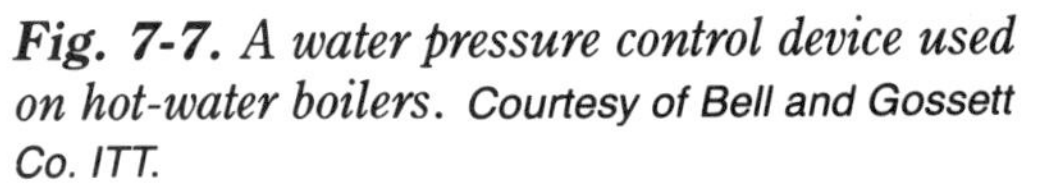

Fig. 7-7. A water pressure control device used on hot-water boilers. Courtesy of Bell and Gossett Co. ITT.

Fig. 7-8. A typical cast-iron boiler section assembled with tapered nipples inserted in the large holes. Courtesy of Weil-McLain Co. A Marley Co.

Removing the Wiring

Removing the wiring to the burner is a simple operation, just shut the power off to this circuit. The electrical circuit supplying power to the burner should not have anything else connected to it. This is to prevent interruption of power to the heating equipment for any reason, such as another appliance blowing the fuse or tripping the circuit breaker. Remove the switch cover and the switch from the switch box. Disconnect the switch wires; one will go to the switch, and the other will go to the burner, both will be black. The white wire might continue through the box or it might be spliced. Remove the conduit and box from the boiler casing. If the upper end of the conduit is not fastened to a

Fig. 7-9. *A cast-iron, hot-water boiler with an induced-draft fan. The motor is shown in the upper-left of the photo.* Courtesy of Weil-McLain Co. A Marley Co.

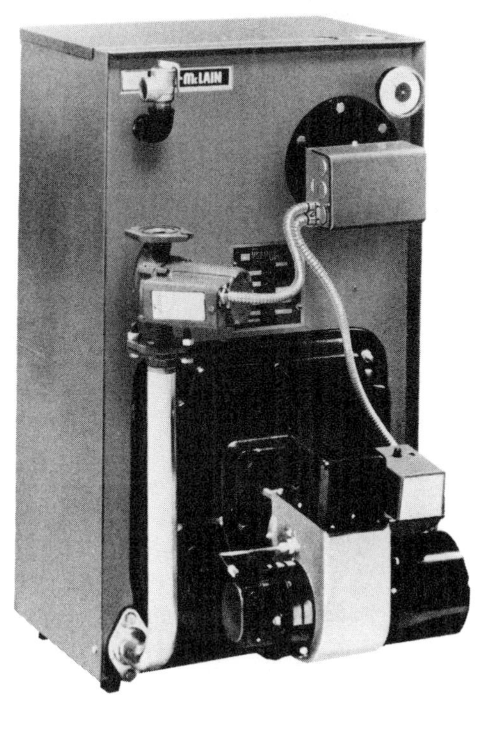

Fig. 7-10. *A completely wired and piped oil-fired, hot-water boiler, ready for quick hook-up.* Courtesy of Weil McLain Co. A Marley Co.

joist, pull the conduit off the Romex cable. Cable is the usual method; enough length is left so that the cable reaches the switch. Some installations have a junction box mounted on a joist and separate wires feed the switch and burner through the conduit. Coil the wires up overhead after first taping or putting wire connectors on each wire end singly. Do *not* connect the wires together. The thermostat wires should also be removed from the boiler casing and coiled near the ceiling. These will be dead.

Removing the Smoke Pipe

Remove and discard the smoke pipe. The new boiler will usually require a smaller smoke pipe. Remove the stack relay if the fuel is oil. This can be re-used. Connect a hose to the boiler drain and drain the boiler. **Caution:** Older boilers were insulated with asbestos. This is a **very dangerous** substance, and **must** be removed by trained technicians. Newer boilers now have glass fiber insulation. See Figs. 7-11 through 7-13, modern boilers.

Fig. 7-11. *A gas-fired, hot-water boiler equipped with an automatic flue damper to save fuel.* Courtesy of HydroTherm Co.

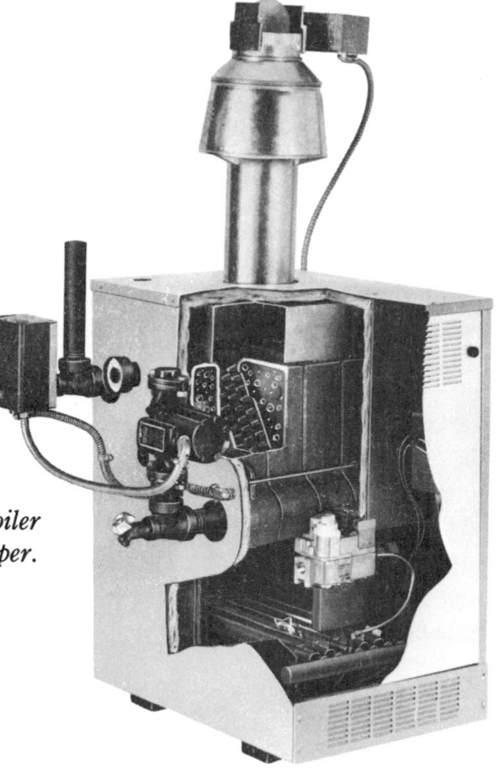

Fig. 7-12. *A gas-fired, hot-water boiler equipped with an automatic flue damper.* Courtesy of Burnham Corp.

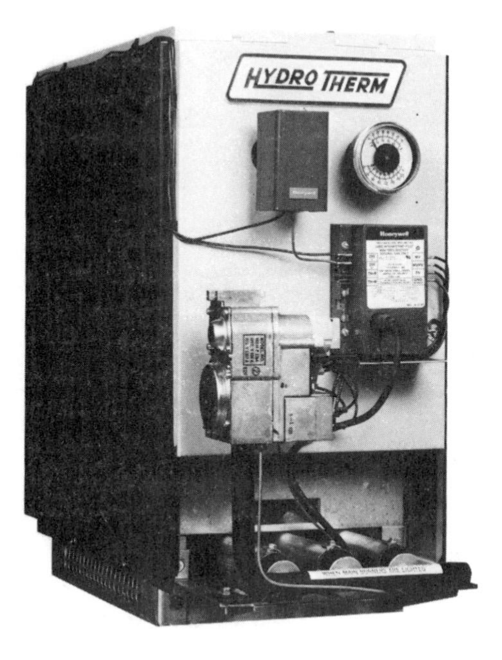

Fig. 7-13. *A gas-fired, hot-water boiler equipped with an intermittent ignition. Notice the ignition control module on the front right.* Courtesy of HydroTherm Co.

Removing the Gas Line

If the boiler is gas fired, the gas line feeding the burner should be removed at least from the ceiling down to the boiler casing. **Be sure** to shut the gas off at the meter before starting to disconnect the line and cap off the open end of the pipe at the ceiling. Save this piping that you have removed to use on the new boiler. Always use two wrenches when working with threaded pipe and fittings, one to turn the pipe and one to hold back on the pipe or fitting that you do not want to loosen or tighten. (See Fig. 2-12.)

Disconnecting the Oil Line

If the boiler is oil fired, shut off the oil line at the tank valve, then disconnect the oil line at the burner. Be sure to protect it from damage while removing the old boiler and installing the new one. The oil line can be plugged by using the fitting from the oil pump and capping the male pipe thread end with a pipe cap. The fitting has a quarter-inch thread, and a quarter-inch pipe plug can then be used to plug the pump opening. **Note**: If natural gas has become available in your area, you might consider converting to this more economical fuel. Your fuel oil supplier can pump the oil from your tank and give you a refund for it.

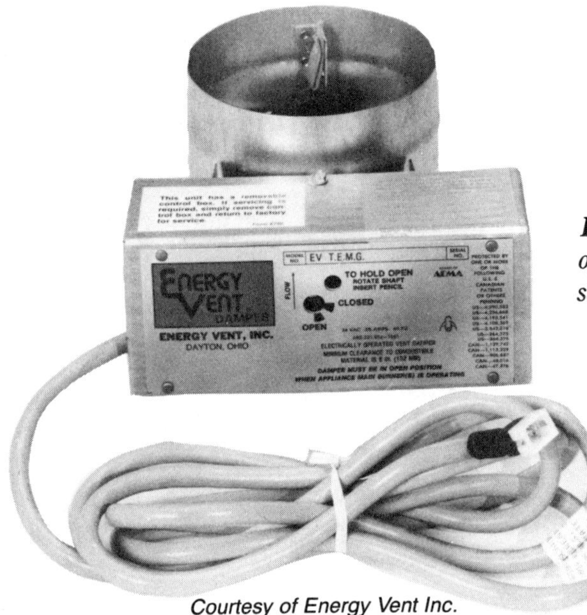

Courtesy of Energy Vent Inc.

Fig. 7-14. *An electrically operated vent damper that saves heat and energy.*

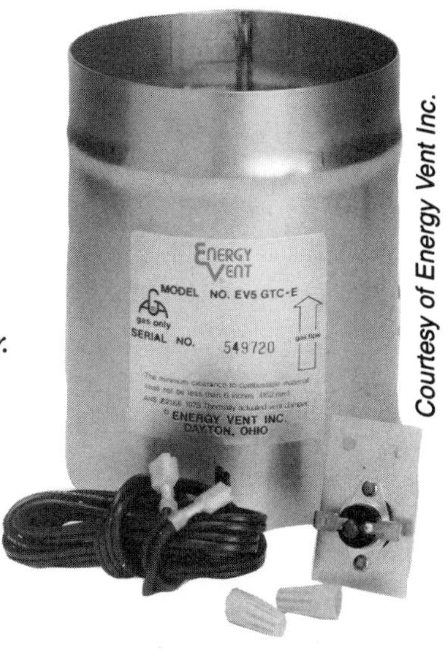

Courtesy of Energy Vent Inc.

Fig. 7-15. *A thermally operated vent damper.*

CONNECTING THE GAS LINE

Earlier in this chapter, I discussed how to remove the gas pipeline so you could remove the boiler. The line will be either three-quarter or one inch size. Gas pipe and fittings are almost always black iron. Shutoff valves are brass. You might have to buy more pipe, depending on the connections. The pipe union is suitable to use again if it did not leak before. The shutoff valve should be mounted about five feet above the floor as before. Do not forget the dirt leg, as this helps to protect the automatic gas valve from sediment and metal particles. Figures 4-11 and 4-12 show the valve and dirt leg. The gas line could be reduced, depending on the gas rating of the burner as it drops down from the ceiling. The reduced size is one-half inch minimum. If the old pipe is three-quarter inch, use it if it is long enough. You might have to reduce the size at the valve if necessary.

Starting at the burner automatic valve, run the piping. Cut and thread a section of pipe long enough to extend two to four inches out beyond the boiler casing to provide clearance. In a vise, screw the side opening of a pipe tee onto the end of the pipe. Be sure to tighten the tee snugly, then start the other end of the pipe into the automatic valve. Be careful that you do not cross-thread this connection, as the valve is of die cast metal, much softer than pipe fittings or brass valves. Be sure to hold back on the automatic valve. Do not tighten excessively because the valve body can split. A 14- or 18-inch pipe wrench is sufficient. As you tighten the tee and pipe assembly, watch the tee and make sure that the tee ends up vertically with the other two openings. A pipe connection cannot be "backed up" (unscrewed slightly), this will cause a leak. After the tee and pipe section are in place, screw a pipe cap on a six-inch nipple using a vice, then screw the other end into the bottom opening of the tee; this is the dirt leg. Again, be careful that no strain is put on the automatic valve of the burner when you are assembling the pipe and fittings. Measure from the face of the tee vertically to a point near the ceiling joists and on the same elevation from the floor as the center of the present gas line. Pipe fittings are measured to the face of the fitting, to the center of the fitting, and to the end of the pipe as shown in Figs. 2-22 and 2-23. The threaded end of a 1/2- or 3/4-inch pipe enters the fitting 1/2 inch. Another 1/2 inch from the end of the pipe end is the center of an el or tee. Both the 1/2- or 3/4-inch pipes have the same number of threads per inch, therefore the dimensions just mentioned are the same. Put a box or other support under the dirt trap to hold the assembly in place. When making up the final assembly of valve and union, remember that coming down from the ceiling (the supply) to the tee, the correct order is: Pipe section, gas shutoff valve, nipple, pipe union, pipe section, and tee. This sequence allows the gas to be shut off to the furnace and the automatic gas valve repaired or replaced with no gas escaping. Due to the smaller or different size of the new boiler, you might need to make an offset in the supply line. To accomplish this, you can make the offset in line coming down from the joists or shorten or lengthen the horizontal line running on the underside of the joists. A change at the ceiling line often looks better and is easy to do. When all the piping has been completely done, you can check for gas leaks in the new piping connections.

Before you begin, make sure all of the pilot lights and open flames are extinguished! *Do not* smoke! Use a thick soap solution or a child's bubble solution. Turn on the gas and check for leaks, using a flashlight. If a threaded joint leaks, a bubble will form (you might also smell gas). The leaking joint will have to be taken apart. Inspect the threads on the pipe and inside the fitting for defects. By cleaning the threads thoroughly before inspecting them, you can discover possible defects. Many times a leak will show up on a section of pipe that cannot be removed as there is no union to disconnect. The only solution is to locate a union, turn off the gas, disconnect the union, and remove pipe and fittings until the leaking section is reached. Then the joint can be reassembled using pipe joint compound. NOTE: A pipe section between two solid fittings (el, tee, or other like type) *cannot* be tightened to stop one end from leaking as the other end will be loosened. The only repair is as described above.

If no leaks show as bubbles, and the gas pressure is still on the piping, a second test for leaks is to observe the gas meter. On the meter face is a small additional dial called "1/2 cubic foot" or "2 cubic foot" dial. If the hand on this dial moves within two minutes, there is still a leak. If the hand does not move, the system is gastight. Both natural and Liquified Petroleum Gas (LPG) have an added chemical that puts off a strong odor so that a gas leak can be noticed immediately.

Hot-Water System Parts

Additional components other than the hot-water boiler are needed, and include: hot-water circulating pump, Figs. 7-16, 7-17; Airtrol fittings, shown in Fig. 7-18; safety relief valves, Figs. 7-19 and 7-20, Monoflo fittings to divert the hot-water flow into the radiation, Figs. 7-21 and 7-21A, and Zone Valves, Fig. 7-22. A cut-away view of a pressure relief valve is shown in Fig. 7-23, and a cut-away view of a pressure relief and a pressure regulator is shown in Fig. 7-24. Figure 7-25 shows a commercial low-water cutoff.

Courtesy of Bell and Gossett Co. ITT.

Fig. 7-16. *Two, hot-water circulating pumps. The one on the left is equipped with a hot-water circulating pump with flexible coupling, shaft packing, a shaft seal, and impeller.*

Wiring the New Gas Burner

If the new burner is a replacement for an obsolete or defective burner, there should be an electrical circuit reserved exclusively for the heating equipment. For a completely new installation, an exclusive circuit should be provided.

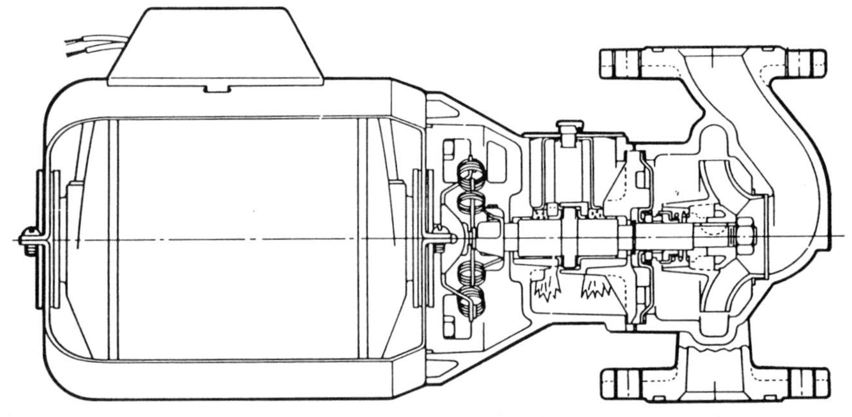

Fig. 7-17. *A circulating pump that can be mounted in vertical pipe lines.* *Courtesy of Bell and Gossett Co. ITT.*

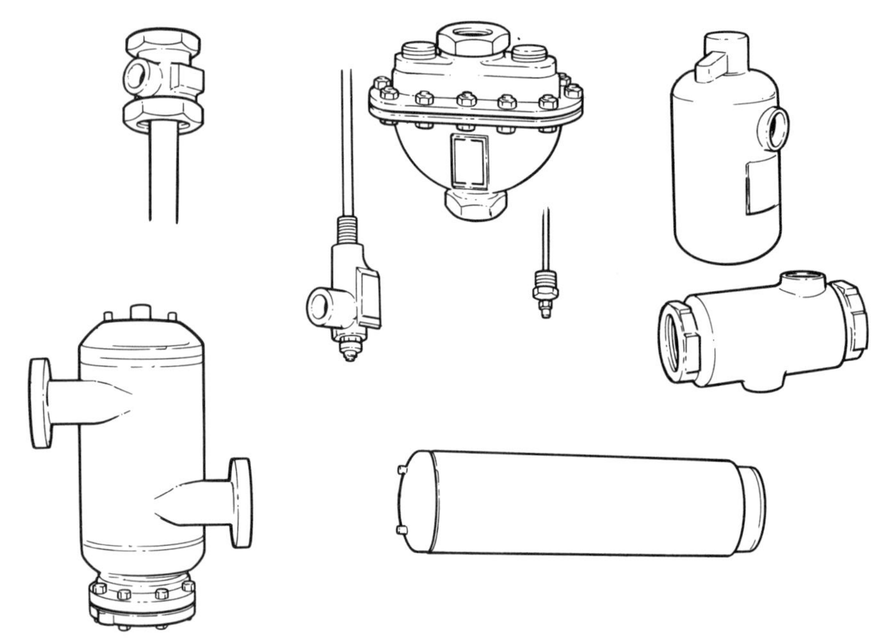

Fig. 7-18. *Boiler trim used on hot-water heating boilers.* *Adapted from a Bell and Gossett Co. ITT drawing. Used with permission.*

This circuit should be wired using 12-2 with ground cable (Romex). This cable should be protected by a 20A fuse or 20A circuit breaker. Run the cable from the electrical panel overhead through holes bored in the centers of the floor joists to the heating equipment. Mount an electrical handy box on the side of the equipment casing. Run a section of half-inch thinwall (EMT) from the side of

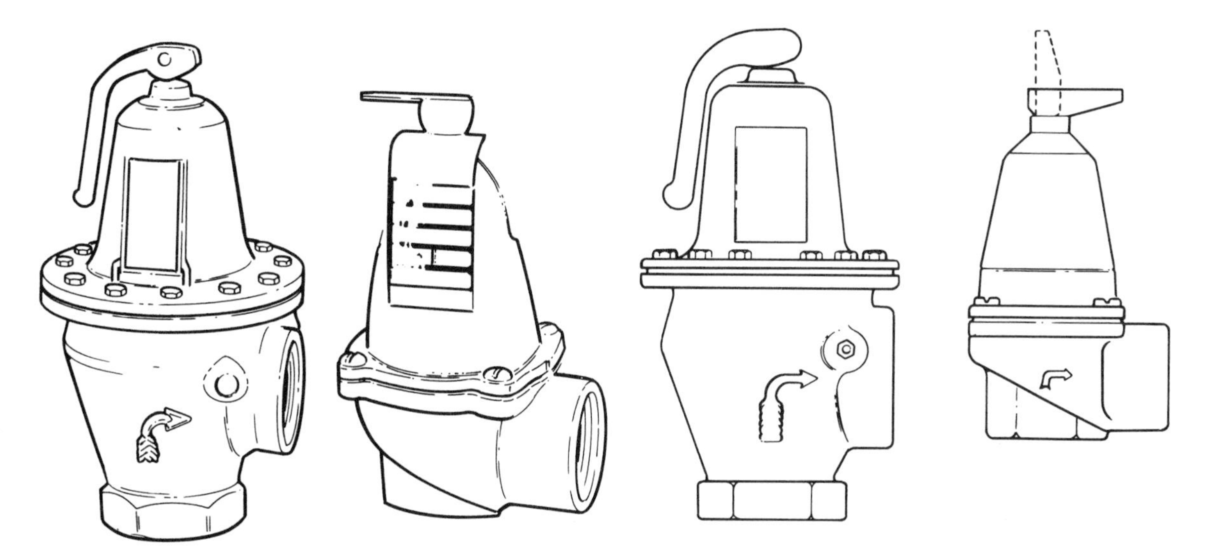

Fig. 7-19. *Two views of a safety relief valve used on fired and unfired pressure vessels that protects against too much pressure.* *Adapted from a Bell and Gossett ITT drawing. Used with permission.*

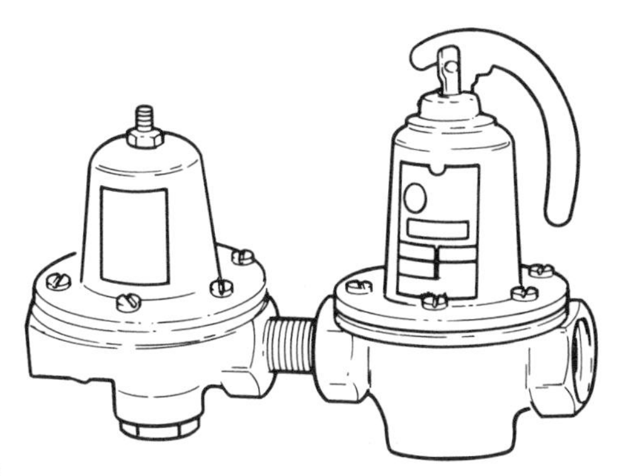

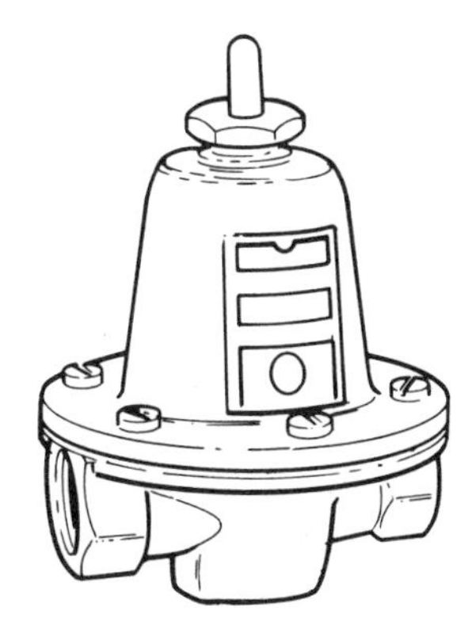

Fig. 7-20. *A combination pressure regulator and safety relief valve.* *Adapted from a Bell and Gossett ITT drawing. Used with permission.*

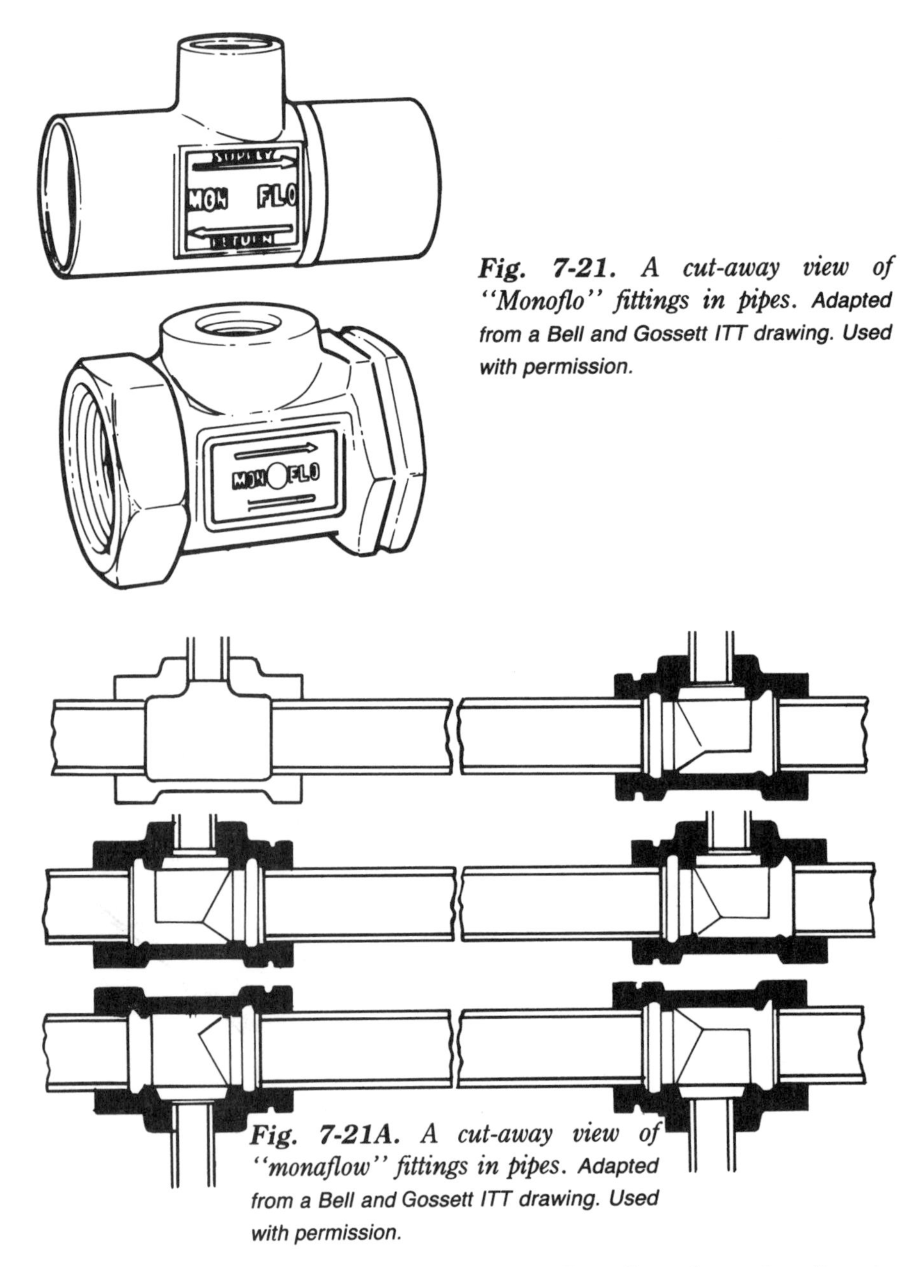

Fig. 7-21. A cut-away view of "Monoflo" fittings in pipes. *Adapted from a Bell and Gossett ITT drawing. Used with permission.*

Fig. 7-21A. A cut-away view of "monaflow" fittings in pipes. *Adapted from a Bell and Gossett ITT drawing. Used with permission.*

a nearby floor joist down to meet the handy box. You will need to make offsets in the thinwall so that it can be anchored to the joist and to the casing. The bare end of the thinwall must have a bushing on the end attached to the joist. This is required by the 1987 National Electrical Code to protect the cable from abrasion. The cable will slide easily down through the thinwall you have installed.

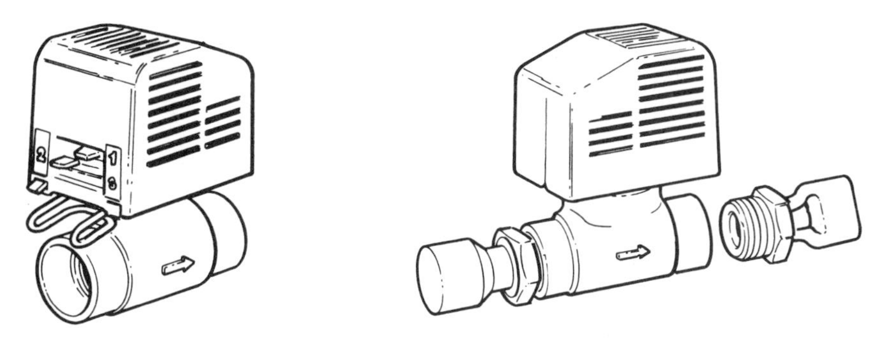

Fig. 7-22. *Zone control valves used on hot-water heating systems, separated into individual zones. Each control valve has its own thermostat. Most valves have an end switch to stop the circulating pump when the valve is closed.* *Courtesy of Bell and Gossett Co. ITT.*

Fig. 7-23. *A cut-away view of a steam boiler pressure-relief valve.* *Courtesy of McDonnell and Miller Co. ITT.*

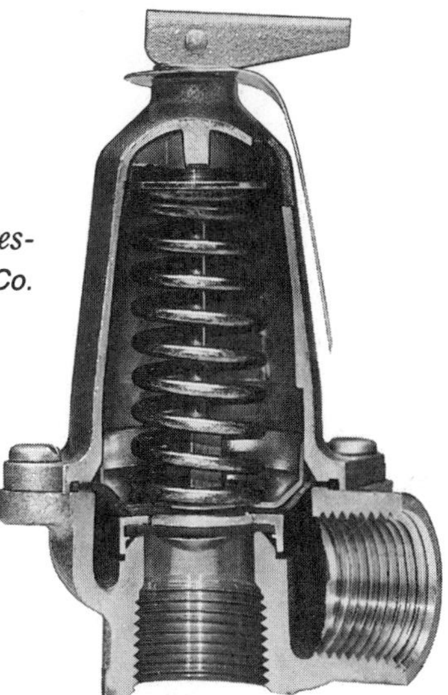

Wire the burner and related controls according to the wiring diagram furnished with the burner and furnace/boiler. The safety and operating controls are furnished and can be attached to the new burner. The burner might have a terminal board, in addition to the wiring diagram that can be attached to the automatic valve body. Refer to Figs. 8-11 and 8-12 for examples of wiring diagrams for gas burners.

Adapting the Gas Burner to the Furnace

Most furnaces or boilers that are gas designed will not accept the so-called conversion burner. This is because of the flue design, which requires the

Courtesy of McDonnell and Miller Co. ITT.

Fig. 7-24. *A cut-away view of a pressure-relief valve, combined with a pressure regulator, that maintains correct pressure in a hot-water heating system.*

Fig. 7-25. *A low-water cut-off for commercial service.* Courtesy of McDonnell and Miller Co. ITT.

Fig. 7-26. *A gas-fired steam boiler with the front casing removed to show the automatic gas valve and the induced-draft fan.* Courtesy of the Burnham Co.

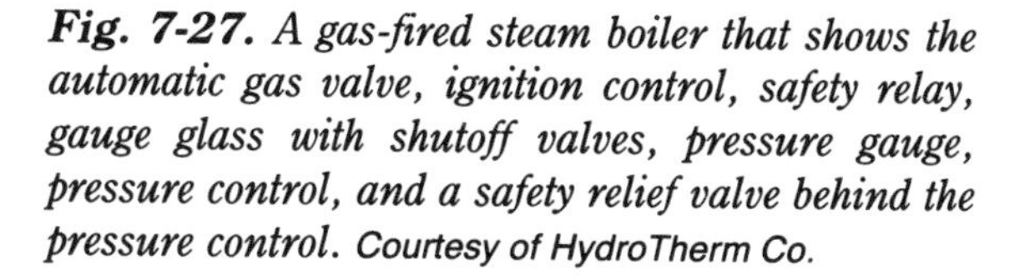

Fig. 7-27. *A gas-fired steam boiler that shows the automatic gas valve, ignition control, safety relay, gauge glass with shutoff valves, pressure gauge, pressure control, and a safety relief valve behind the pressure control.* Courtesy of HydroTherm Co.

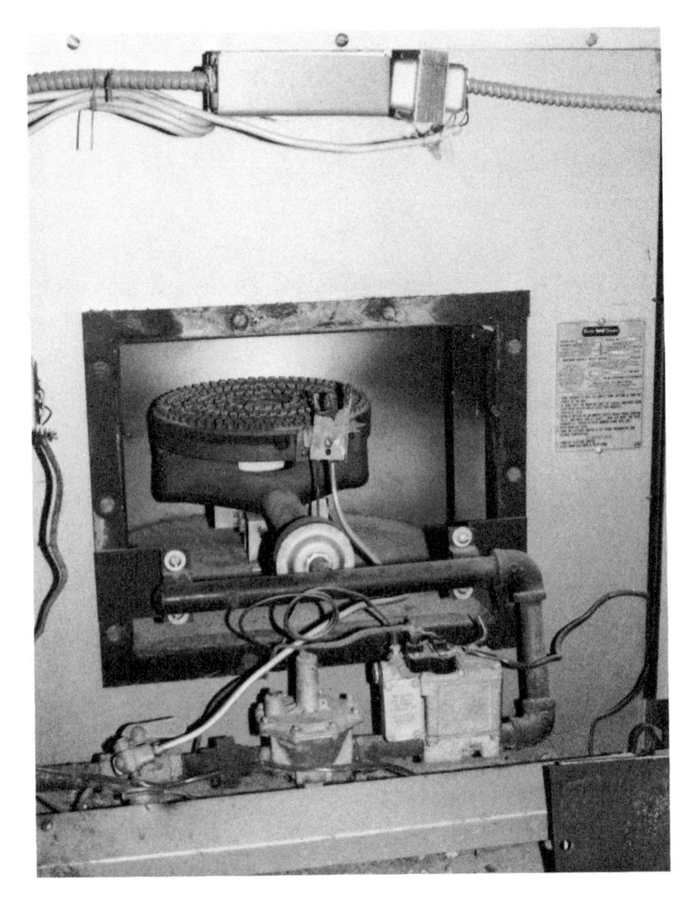

Fig. 7-28. *A older warm-air furnace with a round burner head. The pilot is mounted on the edge of the burner head. This might be a conversion installation.*

gas flame to be directly beneath the flue passage. Be sure to tell the dealer that you must have the exact replacement burner ordered from the manufacturer through the local dealer.

As gas burners have few operating parts, it is often better to replace parts, such as the automatic valve, pilot burner, and possibly, the burner orifices (see Fig. 7-29). The burner heads will need to be replaced only if they are severely damaged or badly corroded so that all the flame characteristics are severely changed. Most replacement parts can be obtained from the dealer handling that brand of burner.

BOILER CONTROLS

Even though boiler controls serve the same function as warm-air furnace controls, they are entirely different. Boiler controls for a hot-water system include a:

1) Boiler operating control, set for Off at 200°F, On at 190°F
2) Boiler limit control, set for Off at 215°F, On at 195°F
3) Boiler fill and relief valve combination, set for 12 pounds
4) Pressure gauge
5) Circulating pump operating control relay (controlled by the room thermostat)
6) Circulating pump
7) Circulating pump low-limit control, set at 125°F. This low-limit control prevents the pump from circulating cold water. It is not necessary, but may be included
8) Low-water cutoff that prevents the burner from operating if the boiler has no water in it

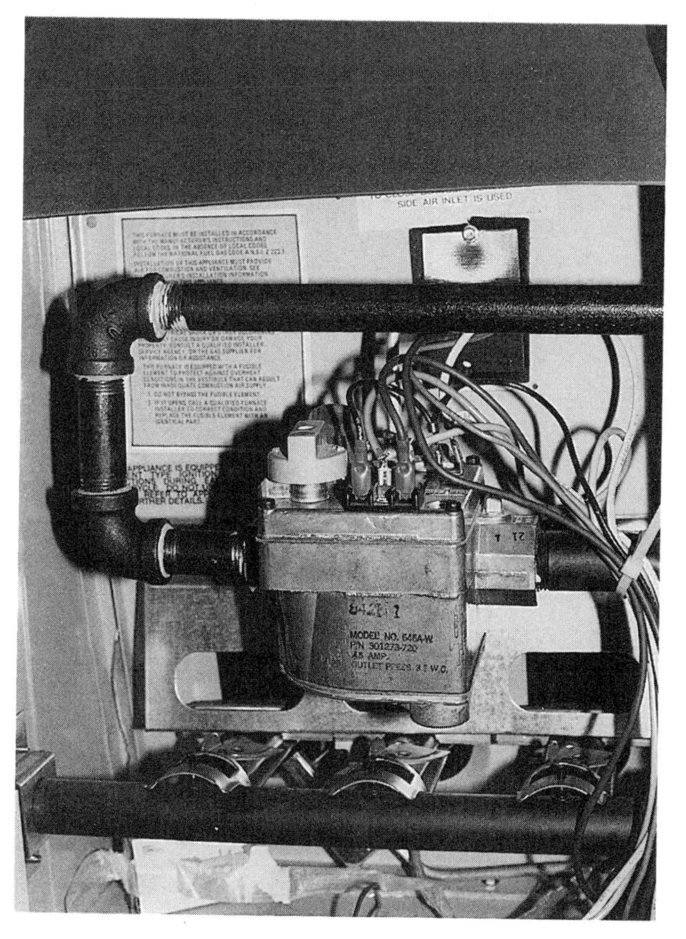

Fig. 7-29. *A conversion installation of an automatic gas valve for a warm-air furnace. The gas supply comes in at the left, enters the valve, then goes to the burner heads below.*

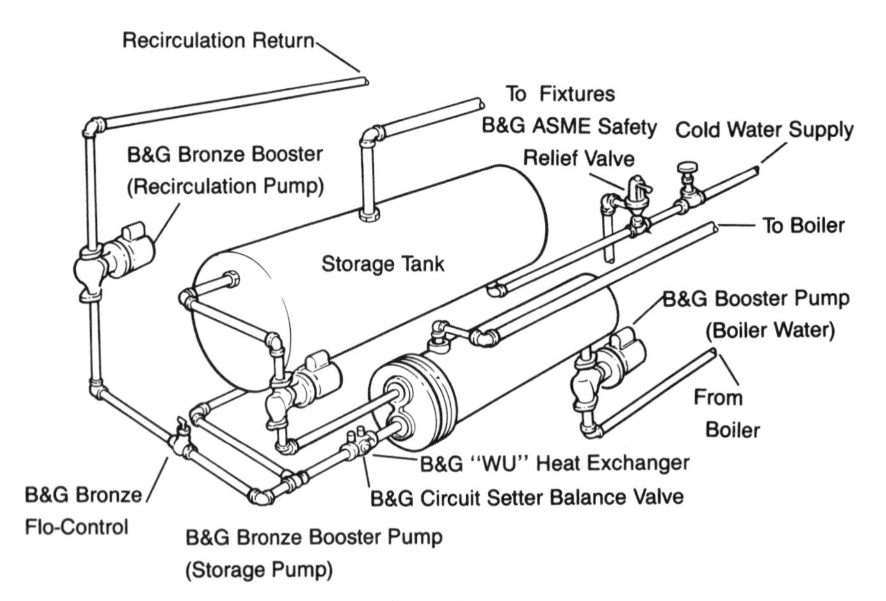

Fig. 7-30. *This water system uses circulating pumps.* Courtesy of Bell and Gossett Co. ITT.

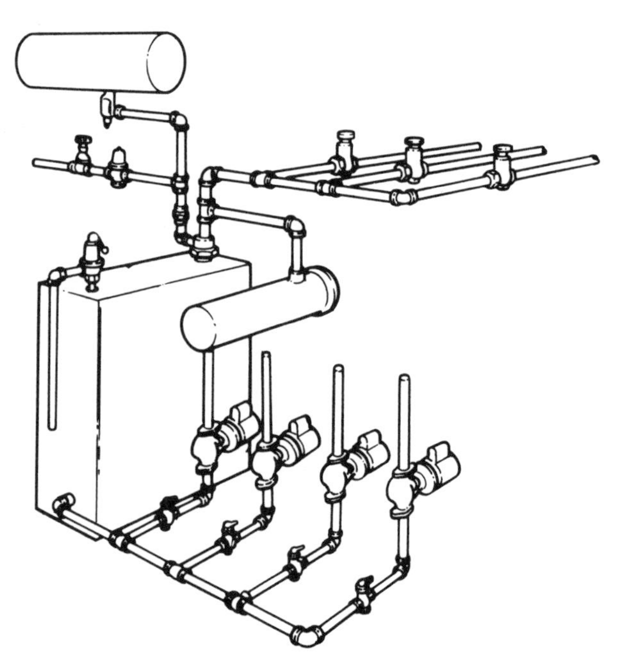

Fig. 7-31. *A multilple zone system. Notice the four pumps that serve the four zones and their related accessories.* Courtesy of Bell and Gossett Co. ITT.

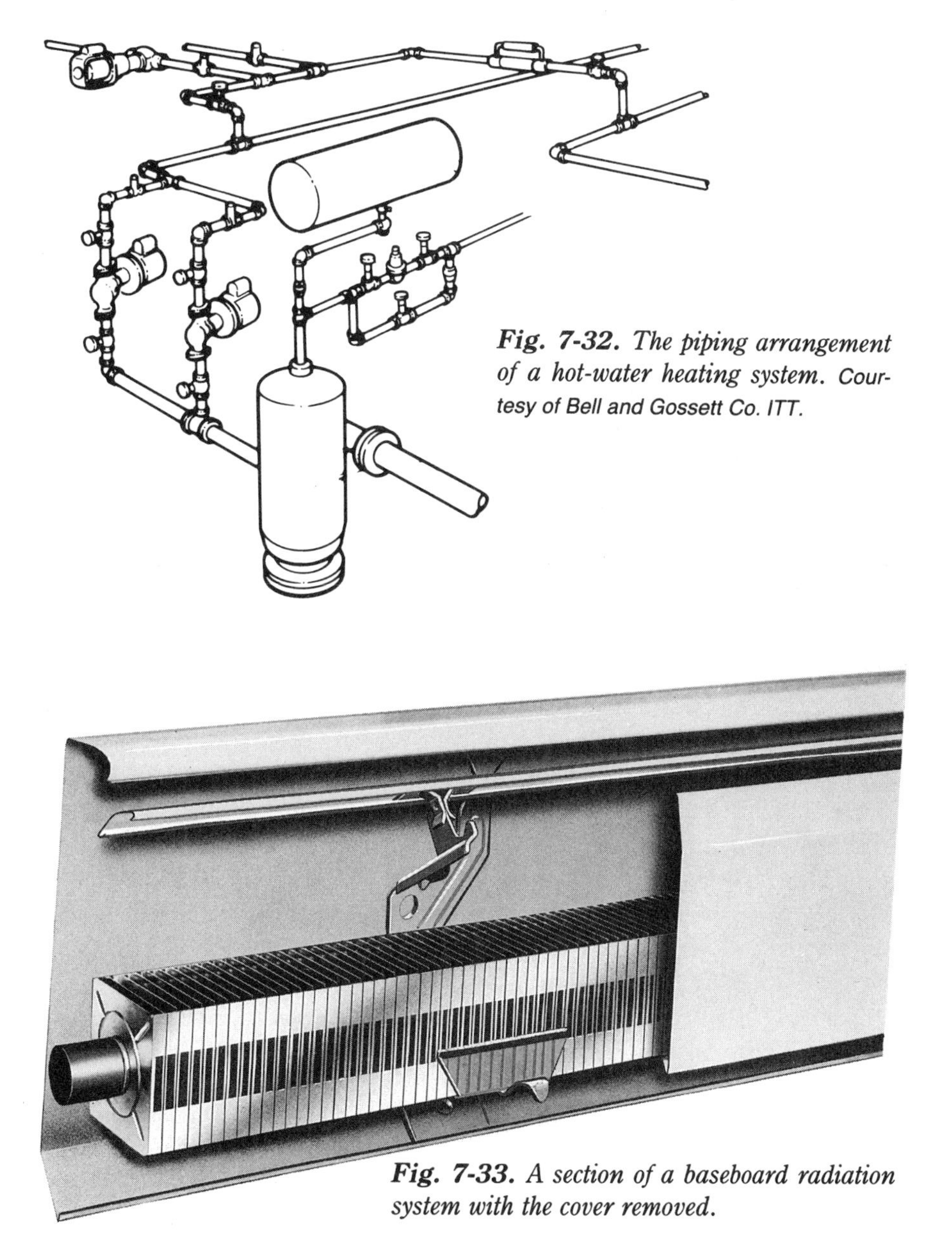

Fig. 7-32. The piping arrangement of a hot-water heating system. Courtesy of Bell and Gossett Co. ITT.

Fig. 7-33. A section of a baseboard radiation system with the cover removed.

Boiler controls for a steam boiler system include: 1) Operating pressure control, set for Off at 2 pounds, On at 1/2 pound; 2) High-limit pressure control, set for Off at 5 pounds; 3) Safety relief valve, set to relieve at 15 pounds; 4) Low-water cutoff in combination with an automatic feed-water valve. The feed-water valve maintains the proper water level in the boiler at all times. This valve has a float that responds to the boiler water level and adds water if needed. A float, similar to a toilet tank float in principle, operates the valve. On the side of a boiler are three small valves (try-cocks) plus a gauge glass. The three try-cocks are screwed directly into the boiler casting in a vertical row. The top and bottom try-cocks are about ten inches apart; the third try-cock is centered midway between. Open the bottom try-cock, if water comes out, the water in the boiler is at, or above, this level. Open the center try-cock, if water comes out, the boiler water level is at, or above, this level. If water comes out from the upper try-cock, the boiler is overfilled and some water must be drained from the boiler.

In normal operation, the water is at the center try-cock. Water might come out, depending on whether the boiler is steaming. This is of no concern unless *no* water comes out of the *bottom* try-cock. In this instance, shut the boiler down and investigate the reason. As an additional check, a gauge glass is installed near, or adjacent to the try-cocks. This glass has two gauge cocks to close off the glass from the boiler for replacement of the glass tube. The glass is heat resistant Pyrex glass, $^5/_8$ inch in diameter and long enough to fit into recesses in each gauge cock. Rubber washers seal the glass from leaking. This gives an additional check on the water level in the boiler.

Residential boilers have *one* low-water cutoff. Many commercial boilers have *two* low-water cutoffs. There have been cases of boilers having two, low-water cutoffs, neither of which stopped the burner in time to save the boiler. Sludge accumulation under the floats of both cutoffs prevented the floats from dropping and shutting off the burner. This emphasizes the need for regular draining and cleaning of a boiler, at least once a year. Commercial boilers must be inspected by the state once a year. Even with good boiler maintenance, the boiler water level should be checked monthly during the heating season. Certain installations, both hot-water and steam, have inserted below the water level a "tankless" heater for domestic hot water. A similar system is shown in Fig. 7-4. These heaters have been used for years and are satisfactory, however, the standard forty-gallon gas *storage tank* water is the best. In areas of very hard water and high iron content, a tankless heater will lime-up and need constant cleaning.

The Hot-Water Heating System

The hot-water heating system takes various forms. Currently, finned baseboard radiation is popular. If the house is fairly large, such a system can be "zoned," or divided into two or more sections, such as living, sleeping, and kitchen areas. Each section, or zone, has its own thermostat and control valve. As mentioned before, the boiler water temperature is maintained by the boiler operating control. An expansion tank is mounted above the boiler and is connected to the main discharge line leaving the boiler. The tank is partially filled with water; the balance is air. A drain valve attached to the bottom of the tank permits you to periodically drain extra water from the tank, as water tends to absorb air and the tank fills with water. Pressure in the system is maintained by an automatic fill valve, which is part of the fill and relief valve assembly. An example of a modern, finned radiation system is shown in Fig. 7-33.

A hot-water circulating pump is in all modern hot-water heating systems. This pump is controlled by a thermostat, or two or more if zoned, through a relay. The relay is necessary because thermostats operate on 24V and the pump operates on 120V. The relay has two terminals, each marked "T." The two wires from the thermostat(s) are attached to these terminals—two from *each* thermostat if more than one thermostat is used. Thus, the thermostats are able to start the pump and receive hot water. If one zone becomes warm enough, the pump might continue to run to supply hot water to the other zones. The zone that is satisfied cannot get any warmer, because the zone valve is closed to that zone. Consequently, any thermostat is able to start the circulating pump. Any thermostat is able to close or modulate the automatic valve that serves its zone only, to control the temperature of that zone. An example of these zone valves are shown and described in Fig. 7-22.

The circulating pump is mounted in a vertical or horizontal position, depending on the style of pump and the manufacturer. A casting contains an impeller with its shaft extending through the pump casting. A shaft seal is located in the casting and surrounds the impeller shaft to prevent water leakage at this point. All parts of these pumps are replaceable and are very easy to obtain. After long use, the seal and flexible coupling might fail and need to be replaced. Figure 7-32 shows zone valves, flo-control valves, and a heat exchanger for domestic hot water. Figure 7-31 shows a special arrangement of booster pumps and flo-control valves.

ADAPTING AN OIL BURNER TO A BOILER

Adapting an oil-burner to an existing boiler is similar to adapting an oil burner to a forced, warm-air furnace.

Boiler Controls

Controls to operate the hot-water boiler differ substantially from those of a furnace. The controls include:

1) The high limit control, set at 212°F. Most commercial and industrial boilers have two, high-limit controls for additional safety, as these boilers operate at much higher pressures.

2) The operating control, usually set at Off at 210°F, On at 200°F.
3) A feed water/pressure relief valve combination, which maintains the correct amount of water in the system and relieves excess pressure when necessary.
4) A low-water cutoff, which stops the burner from operating if the water level in the boiler falls below a predetermined point. This control was formerly not thought necessary, but after numerous boiler failures were reported, it was included as standard equipment. Many times, a boiler would be drained for cleaning, or other reasons, and the burner could still run. This caused major damage to the boiler castings.
5) The pump relay; this relay is actually not part of the boiler controls, as it is controlled by the thermostat.

The Hot-Water Boiler System

The operating cycle of the oil burner/hot-water boiler system is:

1) The boiler water is maintained constantly at 210°F by the operating control. Another control, the high-limit control, is set at 212°F to 214°F. This is the safety control.
2) The circulating pump, under the control of the thermostat, starts and stops on a call for heat. As the boiler water is hot at all times, an increase in room temperature occurs quickly, and shortly thereafter, the thermostat stops the pump. There are sophisticated controls, which, by means of an outdoor sensor, readjust the boiler temperature according to the rise or fall of the outdoor temperature, thus saving some fuel usage. These types of controls are not always necessary, although you might encounter them occasionally and should understand their purpose.

You might come across some hot-water systems called ''gravity hot-water,'' that do not have a circulating pump. These are not satisfactory systems siimply because gravity is the only means of circulating the water through the radiators. You could probably install a circulating pump on this type of system for improved operation, and more even heat in the rooms. Because this type of system is very old, it would be best to replace the radiators with modern baseboard radiation.

Home boilers do not require inspection, although the boiler should be drained and inspected by the owner. Sometimes boiler treatment is needed or recommended to be used in hard water areas. If you experience problems with your boiler, you can probably call the inspection department and have your boiler inspected and remedies suggested. There might be a fee charged, but this service can help you to take better care of your boiler. Inspectors know local water conditions and can suggest the best treatment for your boiler.

The Steam Boiler System

The steam boiler is controlled directly by the thermostat, and there is a longer On-Off cycle because steam has to be produced each time there is a call for heat. Many older homes have this type of heat, and the owners are usually satisfied. Steam systems have proven satisfactory for many years when well maintained, but modern baseboard radiation is more convenient than large, cast-iron radiators and looks more modern. Radiation also take up less room and space is saved. A modern, hot-water system can be separated into zones, using motorized valves to turn individual zones Off or On. Each of these valves has an ''end switch'' that turns the circulating pump on when the valve starts to open after a call for heat from the thermostat is received. The end switches are wired parallel, meaning that *any* end switch is able to start the pump. Conversely, *all* end switches must be in the Off position for the pump to stop. Moderate size homes may have a maximum of three zones: the living area, sleeping area, and recreation area.

Unfortunately, hot-water- and steam-heated buildings do not lend themselves to central air conditioning. Ductwork would have to be installed at great expense, as opposed to forced, warm-air systems that have the ductwork already in place. Carefully consider the limitation of hot water or steam heating before making a decision, and weigh it against additional cost of installing ductwork for central air conditioning. Window air conditioning units are an alternative, but have the disadvantage of partially blocking the window in which they are installed and often have to be removed at the end of the warm season and stored. Refer to the section on window air conditioners in Chapter 8 for details on installing and servicing these units.

8

Installing Central Air Conditioning

TODAY, EXCEPT IN AREAS WHERE TEMPERATURES ARE VERY MILD, AIR CONDItioning is considered necessary. Most large commercial buildings using air conditioning have sealed windows and natural ventilation is not possible. Single homes, condominiums, and apartments have windows that open, but if air conditioning is available, they are normally kept closed. Some apartment buildings are completely air conditioned, and the windows are sealed. One advantage of this policy is that drapes, furnishings, and other items remain cleaner and require less maintenance. This tends to give the units a better appearance and allows a longer life for the furnishings.

FEASIBILITY

There are times when central air conditioning is either not feasible or is very expensive to install (see Table 8-1). Many older homes are heated either by hot water or steam boilers. Obviously, these types of heating systems have no ductwork. Air conditioning requires ductwork and a blower system. It would be very expensive to install ductwork and a blower system in addition to installing cooling, about twice the cost of an add-on system. Not only would ductwork have to be installed and openings cut for the supply and return registers. You would also need to house both the blower and motor, and the cooling coil. A

Table 8-1

Heating Costs:	
Furnace	$ 366/year
Heat pump	$ 645/year
Cooling Costs:	
A/C unit	$ 219/year
Heat pump	$ 242/year
Installation Costs:	
Gas furnace	$1400
Electric A/C	$1600
Heat pump	$3200
Equipment Life:	
Gas furnace	18 years
Electric A/C	14 years
Heat pump	8 years
Annual Service Contract:	
Gas system	$ 150
Heat pump	$ 200

Annual Equipment Costs:

Gas furnace: $\frac{\$1400}{18 \text{ yrs.}} = \$\ 78$

Electric A/C: $\frac{\$1600}{14 \text{ yrs.}} = \$114$

Heat pump: $\frac{\$3200}{8 \text{ yrs.}} = \$400$

The annual costs for each alternative can be summed and multiplied by the number of years in the study.

	Gas System	Heat Pump
Equipment	$192 (78 plus 114)	$400
Heating	366	645
Cooling	219	242
Maintenance	150	200
Totals	$927	$1,487

15-year cost:

Gas System 15 × $ 927 = $13,905

Heat Pump 15 × $1487 = $23,905

Courtesy of American Gas Association.

house that is built on a concrete slab must have the ductwork installed in the attic and insulated. The original heating system might have the pipes buried in the concrete floor (radiant heat). Although this system usually works well, if a leak develops, it is very expensive to repair. If you have this type of heating system, it might be feasible to change to a forced, warm-air system with air conditioning. There are many options to choose from and only you can decide what is the best system for you. Many small homes use window air conditioners and are very comfortable. Larger window air conditioners or through-the-wall type air conditioners can be installed in an attic vent dormer and ducted to the living room with the noise of the unit muffled because of the attic location.

Before you purchase or begin to install an air conditioning system, you need to consider the size of the house, and the amount of insulation in the walls and ceiling. Newer houses usually have three and a half inches of wall insulation and six inches of ceiling insulation. It is advisable to add another six inches on top of the present ceiling insulation. This can be blown in, roll, or bat type. If roll or bat insulation is used, lay it crosswise on the ceiling joists and ''unfaced'' (neither side has Kraft paper). Side walls of the house are not as important as the ceiling, because most of the heat loss or gain is through the ceiling. Side walls are very difficult to insulate in a finished house.

The area of the country you live in also has some effect on the capacity you will need. Areas with high humidity need a larger capacity unit to remove both the moisture in the air and the heat. Moisture, in air conditioning terms, is called latent heat (hidden), because it cannot be measured with a thermometer. In actual operation, the air conditioner removes this moisture in the form of water (condensate). As this moisture is removed from the air, there is no reduction in the actual temperature, but the area will feel more comfortable because the air is drier. After the humidity has been reduced, the system starts to reduce the temperature until the setting of the thermostat is reached. Very dry areas such as Arizona need only enough capacity to reduce the air temperature. Many southern states and resort areas are subject to high humidity. The midwest and central states usually have humidity in the range of about 40 to 50 percent. During rainy weather, humidity increases.

The normal size house usually requires a 2 1/2 ton (30,000 BTU) system for a 1600 square foot floor area. Basements are not cooled unless there is a recreation area that is used frequently.

Most houses built on a concrete slab have the heating and cooling ducts in the attic where the air is distributed through ceiling registers. It is most important to have these ducts covered with insulation. Insulation installed by the builder is usually just one-inch thick. An easy way to add more insulation is to use six-inch insulation, split flatwise to make a three-inch thick bat. Wrap a length of this three-inch insulation crosswise on the duct and tuck underneath each side to hold it in place. Ductwork with additional insulation plus added ceiling insulation greatly reduces both heating and cooling costs. Modern storm windows and doors should be installed if there are none.

AIR CONDITIONING SYSTEMS

Once the house is well insulated, it is time to choose a system. Carefully consider the room sizes, ceiling heights, and number of windows and doors. Room sizes (floor areas) multiplied by the ceiling heights is the cubic volume of each room. Add these together to get the total volume for the house. You can then give this figure to either an air conditioning contractor or a major store or home center that sells air conditioning systems for the homeowner. These places usually have knowledgeable people who can determine the correct air conditioning capacity you need. The system's capacity must equal the heat gain of the house within close limits, otherwise the unit will "short cycle" (start and stop frequently) and damage the compressor if the unit is oversized. An undersized unit will run constantly because it does not have the capacity to cool the house.

Generally, residential central air conditioning systems are the electric refrigeration type, which is identical to the refrigeration system of the standard, household electric refrigerator or freezer. The difference is that the air conditioner operates at a higher temperature so that cool air is supplied to the living areas.

Another type of refrigeration system known as the absorption refrigeration cycle was quite popular about twenty years ago. This system, using ammonia as the refrigerant, is quite complicated and is difficult to service. The local gas utility sold and serviced these systems. Many years before this system was popular, household gas refrigerators were available. At one time, kerosene refrigerators were also manufactured. All of these appliances used a flame to operate their refrigeration cycle. Technical information and servicing of these older systems are not covered in this book, only the electric refrigeration system is covered.

No matter what type of system you choose, when you receive the air conditioner you purchased, check each component listed on the bill of lading. Occasional errors happen, so check part numbers and sizes, and notify the shipper or any errors. At times, the tubing length might be short. Too long tubing is better, unless the length is extremely long. Four to six feet extra can be coiled in a thirty-inch diameter circle if necessary, either a longer or shorter length must be exchanged for the proper length.

INSTALLATION

Most residential air conditioning systems are "split" systems, which consists of two main components: The cooling coil (evaporator) in the furnace plenum, which is above the furnace cabinet in the sheet metal duct, or in a specially built coil and cabinet that fits directly onto the furnace discharge connection. The second component, the condensing unit, is located outside. The preferred location is on the north or east side of the house because the afternoon and evening sun does not heat the condensing unit, and thus the unit is more efficient.

Refrigerant Tubing

The cooling coil and the condensing unit (Figs. 8-1 through 8-6) are connected together using measured lengths of charged and sealed refrigerant tubing. Two lines are always used, one to carry the refrigerant liquid (the liquid line) and one to carry refrigerant gas (the suction line) (see Fig. 8-7). The suction line is the larger of the two. The ends of both tubes have special sealing couplings that match couplings on the condensing unit and on the cooling coil inside the house. The couplings have a spring-loaded valve in the end of the line that is pressed open when the couplings are connected together. The condensing unit and the cooling coil are both fully charged with refrigerant, so that when the connections to the tubing lines are made, the system is completely charged with refrigerant. These charged and sealed lines are available in various lengths to equal the distance from the condensing unit to the cooling coil in the furnace.

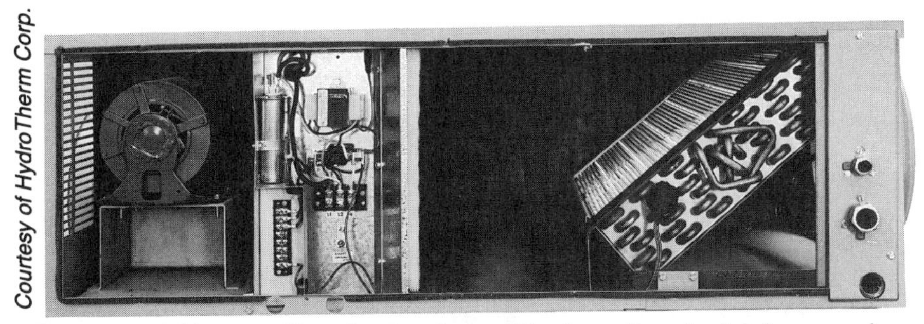

Courtesy of HydroTherm Corp.

***Fig. 8-1.** A blower cabinet for the air conditioning coil and related accessories.*

You will also need enough tubing to allow for bends and offsets. I usually coil the excess length in a thirty-inch loop near the cooling coil and against the basement ceiling joists. Tubing lengths usually increase in five foot increments, from ten feet to perhaps forty feet. See Figs. 8-1, 8-2, and 8-3. The tubing for the suction line is large, $^{7}/_{8}$ inch or $1^{1}/_{8}$ inch outside diameter (OD). If you bend this

Fig. 8-2. An add-on cabinet with an "A"-shaped air conditioning coil that is set on top of the furnace discharge opening. Courtesy of Carrier Corp.

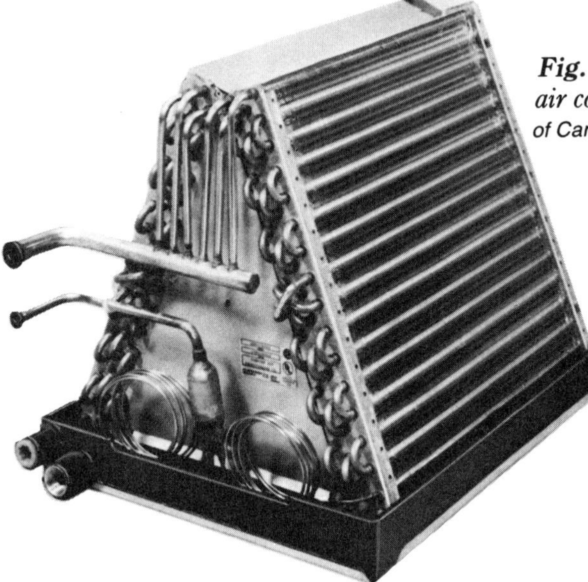

Fig. 8-3. A typical A-shaped air conditioning coil. Courtesy of Carrier Corp.

size tubing to a smaller radius than recommended, a kink will develop, which can damage the tubing and cause a leak. The smallest radius that is recommended is twelve inches. The liquid line is $^{1}/_{4}$ inch or $^{3}/_{8}$ inch OD. The smaller tubing and the larger tubing are routed as a pair and bent to the same radii and go through the same opening into the basement. If the tubing is damaged, it is best to buy a new length. Repairing a charged tubing is more expensive than the cost of a new, pre-charged length. See Fig. 8-3.

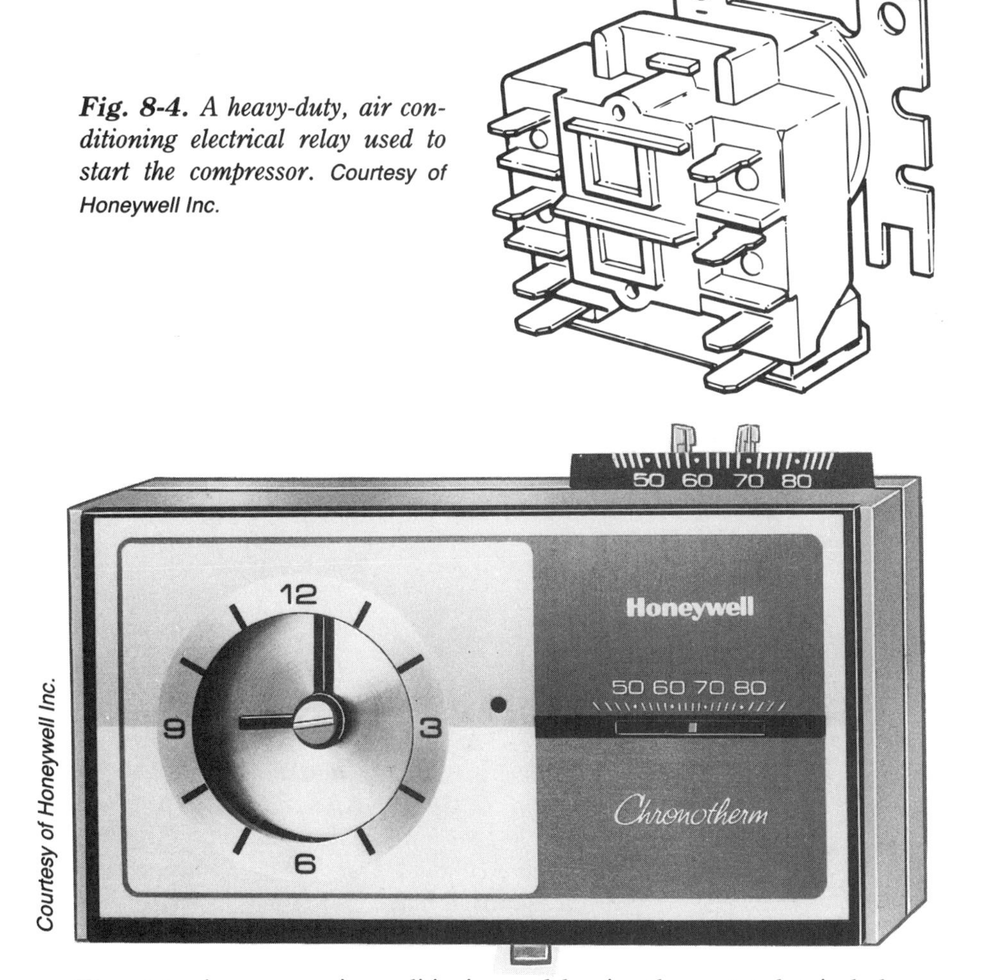

Fig. 8-4. *A heavy-duty, air conditioning electrical relay used to start the compressor.* Courtesy of Honeywell Inc.

Courtesy of Honeywell Inc.

Fig. 8-5. *A common air conditioning and heating thermostat that includes a night setback and cooling set-up.*

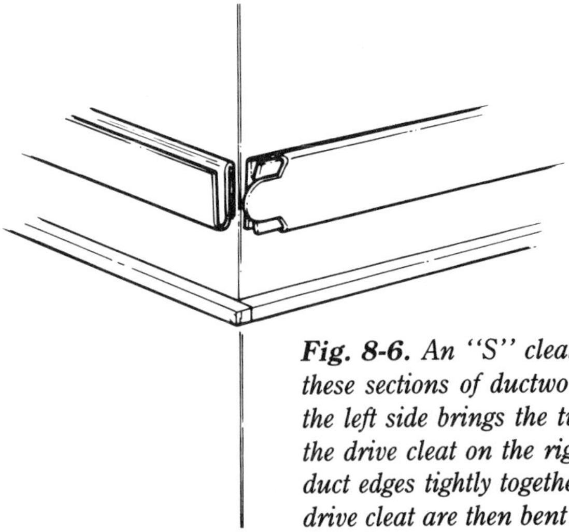

Fig. 8-6. An "S" cleat and a drive cleat connects these sections of ductwork together. The S cleat on the left side brings the two duct edges together, and the drive cleat on the right side draws the other two duct edges tightly together. The rounded ends of the drive cleat are then bent over at the four corners.

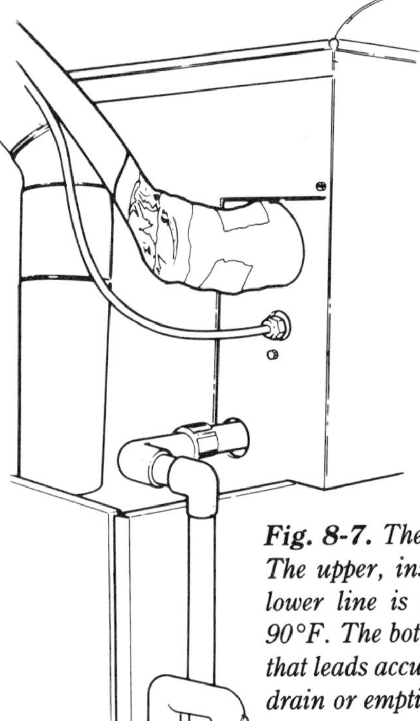

Fig. 8-7. The refrigerant lines to the cooling coil. The upper, insulated suction line stays cold. The lower line is the liquid line and stays at about 90°F. The bottom line is the condensate drain line that leads accumulated condensate to a nearby floor drain or empties into a lift pump.

Before you cut any openings through the basement wall for the refrigerant tubing, examine the basement ceiling for a clear route. First run a fifty foot length of clothesline, and string it along the route proposed for the tubing. This way, you can determine the length of tubing you'll need. Make bends and offsets just as if the line were the tubing. You can ask ahead of time, before purchasing the system, what length of tubing is stocked, then order the correct length. The opening in the wall where the tubing will go should be large enough for the couplings, which are larger than the tubing.

If you have purchased a system that has the coil already installed in a housing made specifically for your make of furnace, then you can disconnect and remove the present plenum (the sheet metal housing just above the furnace casing). The best way to do this is to support the trunk line leaving the plenum with wire or other means, then find, on opposite sides of the plenum, two drive cleats. These are sheet metal strips, three inches longer than the side of the plenum, see Fig. 8-6. The strip is two inches wide and both sides are bent completely over on themselves—but not meeting—to leave a half-inch space down the center. The ductwork edge is also turned back on itself about 3/8 inch. The turned edge on the plenum, and the next section of ductwork, are caught and held by the mating-turned edges of the drive cleat, thus joining the sections. The two other sides of the plenum, or duct, are joined by S-cleats, actually they are "Z" shaped, with the Z nearly closed, leaving enough space for the duct ends to fit into both sides of the Z. On large, commercial installations, these cleats have one side at a 90 degree angle to the flat of the duct. This provides a stiffener that supports the duct rigidly so that it will not vibrate.

Bend the ends of the drive cleats over and around the corner of the duct, and straighten out the end that will allow you to drive the cleat off the duct sections with no interference at the opposite end. A wall or other obstruction might be in the way and prevent you from completely removing the cleat. Drive the cleat flush with the duct corner, then use a large screwdriver or cold chisel to finish driving the cleat off. You will also have to drive the S-cleats off the other two sides of the duct. These cleats have no bent-over ends and are easy to remove. When all four cleats are removed from each end of the section, you can remove the section and set it aside. Do not discard it because you might need it as a partial "fill-in" if the new air conditioning section holding the cooling coil is shorter than the plenum section you removed. For a more workmanlike installation, you might want to have a sheet metal shop fabricate a new, shorter plenum or shorten the present plenum. In most cases, the new cooling section is attached to the furnace with cleats.

At the top of the cooling section a fire-resistant "flexible connection" should be installed to isolate vibration from the plenum and related ductwork. This is standard procedure in all fan-forced, warm-air systems.

By taking careful measurements of the space between the new cooling section top, and the end of the original ductwork, you might be able to have a sheet metal shop fabricate the shortened plenum without having someone come out and take measurements. Make a sketch of the space where the new section will fit. Give all dimensions possible so that the new section will fit properly. Do not forget to have a new flexible connection made to isolate the cooling section and plenum. I have had this sheet metal work made, and if the dimensions are accurate, the new section will fit perfectly.

THE CONDENSING UNIT

While you are waiting to receive your system, you can build the concrete slab to support the condensing unit. You can also install the electrical wiring up to three feet of the condensing unit connection. See the next section on electrical connections in this chapter.

The condensing unit should be mounted on a concrete pad close to the house, preferably on the north or east side. If you will be putting the unit close to the house foundation, allow for the tubing to be bent in a large, gradual curve. If the entrance to the basement is directly behind the condensing unit, or slightly offset, leave enough room for the unit to be serviced.

Wiring the Condensing Unit

Most newer homes have enough electric capacity to supply a two and a half ton (30,000 BTUs) air conditioning unit. Check the service entrance panel for the house. Carefully remove the cover of the panel; do *not* let the panel covers fall back into the live parts inside. There is usually an unused space for a circuit breaker or unused fuse opening. You must have 240 volts available for central air conditioners. Therefore, both hot wires and the neutral wire will be used. You should have no problems doing this wiring. You can refer to my other book *Mastering Household Electrical Wiring*, also by TAB BOOKS, No. 2987 if you need further information on wiring. The book has many illustrations showing how to wire an air conditioning condensing unit. In locations where the electric meter is near the condensing unit, you can run a line from the meter box to a fused disconnect mounted within five feet of the unit cabinet. This work has to be very carefully done, and you might want to hire a licensed electrician to run the line to the disconnect, then you can make the final connection to the condensing unit. To do this work, the electric meter has to be removed, the wiring installed, and the meter replaced. The fused disconnect needs to be the waterproof type. Most have a pull-out fuse block. Fuses are the 30 amp cartridge type, time-delay style.

Electrical Connections

Electrical power for the condensing unit is 240 volts, single phase, 60 hertz, protected by 30 amp fuses or circuit breakers. **Note**: Many air conditioning units state on their nameplate that fuses must be used even though circuit breakers protect the wiring to the air conditioning unit. This is required according to the 1990 National Electric Code, Part C of Article 440, Air Conditioning and Refrigeration Equipment.

As I mentioned in the previous section, you can install the electric wiring up to three feet of the condensing unit connection while you wait for your system to arrive. You will have to install a fused disconnect, 30 amps, to protect the No. 10 wire used. If you are familiar with working with electricity, you can do the wiring yourself. Wiring materials are available at hardware and home center stores, and there are usually people there who can advise you if you are unsure about something. If you are not familiar with electrical wiring, then it would be best to hire a professional. See Fig. 8-11.

There is no electrical connection to the cooling coil because the furnace fan blows the air through the coil. (Figure 8-4 shows a fan relay for air conditioning.) An existing furnace retrofitted for air conditioning might need to have the furnace fan replaced with a larger or higher speed fan. If you need to buy a new fan for a replacement, I recommend a three-speed. They can be adjusted for proper airflow. Greater airflow is needed because air conditioning requires more air to pass through the cooling coil than for heating only.

The Cooling Coil and Related Plumbing

Cooling coils are made in two styles: the "A" shape, and the flat coil. These are shown in Figs. 8-1 and 8-3. The A shape is more common. It provides the same capacity, but takes up less vertical height. Height can sometimes be a critical dimension because of a low ceiling in a basement or other area. The A shape is two slanting coils joined at the top and held together by plates at each end. All cooling coils, when operating, form condensate (water) that must be removed. The coil is mounted in a drain pan about two inches deep. The pan has a drain connection threaded for 3/4 inch or 1/2 inch pipe threads. A drain line must be connected to the pan drain outlet using either hose, copper, or plastic. Many times, there is a floor drain nearby, usually so that the water heater can be flushed. There are cases where the floor drain is some distance away or beyond a traffic area. In this case, you will need to install a small lift pump and raise the drain line up to the ceiling and over and down to the drain location. These pumps cost about $50, but are the best way to dispose of the condensate. On very humid days, a great amount of condensate is collected (the coil sweats as a glass of iced beverage does). The small condensate pump operates automatically. A separate pan is provided to hold the condensate pump and collect the condensate for the pump. Most pumps include a pan, but some do not. You can run a drain line from the coil condensate pan to the condensate pump, which can be set on the floor or mounted on the side of the furnace about twelve inches *below* the level of the coil pan. You will also have to provide an outlet for power to feed the pump (120 volts, 60 Hz). If no outlet is close by, it is easy to extend a line from an uninterrupted source, such as a pull-chain socket or wall receptable. Cooling lines and a condensate drain are shown in Fig. 8-7.

The Furnace Fan and Motor

Most heating-only furnaces have a single-speed motor on the fan. This is not sufficient for air conditioning, because more air volume must pass through the cooling coil. The temperature difference between entering air and leaving air through the coil is only 15° to 20°, therefore more volume is needed. In the heating mode, the temperature difference is about 70°, thus the volume of air through the furnace is much less.

Because of the increased air volume, a two- or three-speed fan should be installed. Only *one* speed setting is used to achieve optimum results. In very rare situations, the furnace ductwork is found to be undersized to the point that it must be replaced with larger ductwork.

THERMOSTATS

More than likely, you will need a new thermostat because heating thermostats do not have the connections for cooling (see Fig. 13-3). Some thermostats are wired for both heating *and* cooling and need only the addition of a subbase. The subbase is mounted on the wall first, then the thermostat is mounted on the subbase with captive screws, which screw into threaded holes in the subbase. On the subbase are two levers labelled: *Heating—Off—Cooling*, and *Fan—Off—Auto*.

A furnace designed for air conditioning will already be fitted with space for the cooling coil and have a multispeed fan installed. Many furnace manufacturers also make air conditioning systems suitable for installation in their brand of furnace. The cooling coil is designed to fit exactly on top of the furnace connection to the ductwork. If this is the case, the plenum will be shorter than normal to allow for the special section (about eighteen inches high). Beyond the duct extension will be the trunk line or lines to distribute the air properly. Contact a local distributor of your make of furnace and you will be able to find the cost and availability of the add-on equipment. See Figs. 8-6, 8-9 and 8-10.

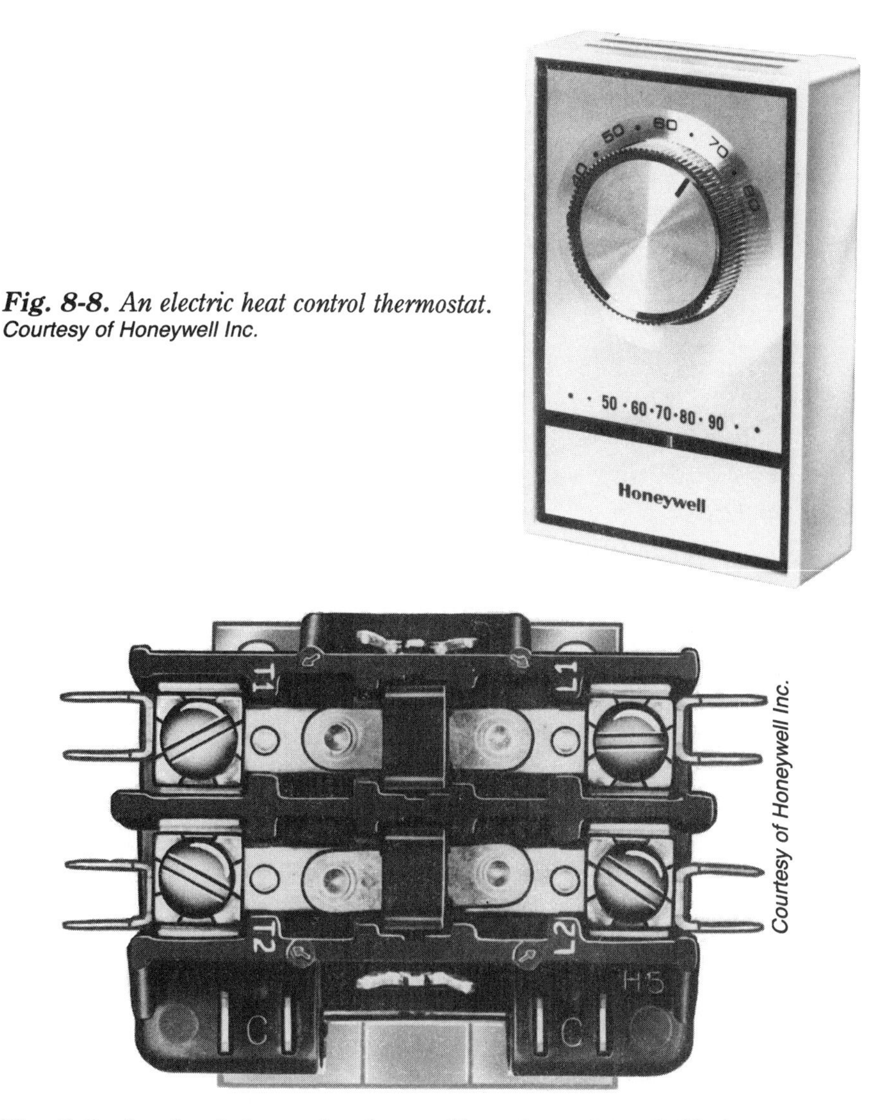

Fig. 8-8. *An electric heat control thermostat.*
Courtesy of Honeywell Inc.

Courtesy of Honeywell Inc.

Fig. 8-9. *An electric heat relay that enables a low-voltage (24V) thermostat to control the line voltage (120/240V) electric heat.*

Wiring the Thermostat

The thermostat for the air conditioning and the heating systems can be either a standard heating/cooling thermostat, or a heating/cooling night set-back type (see Fig. 8-5). Night set-up types allow the cooling temperature to be raised during the night. This type for raising the cooling temperature is used infrequently. During hot summer nights cooling is needed and this feature is

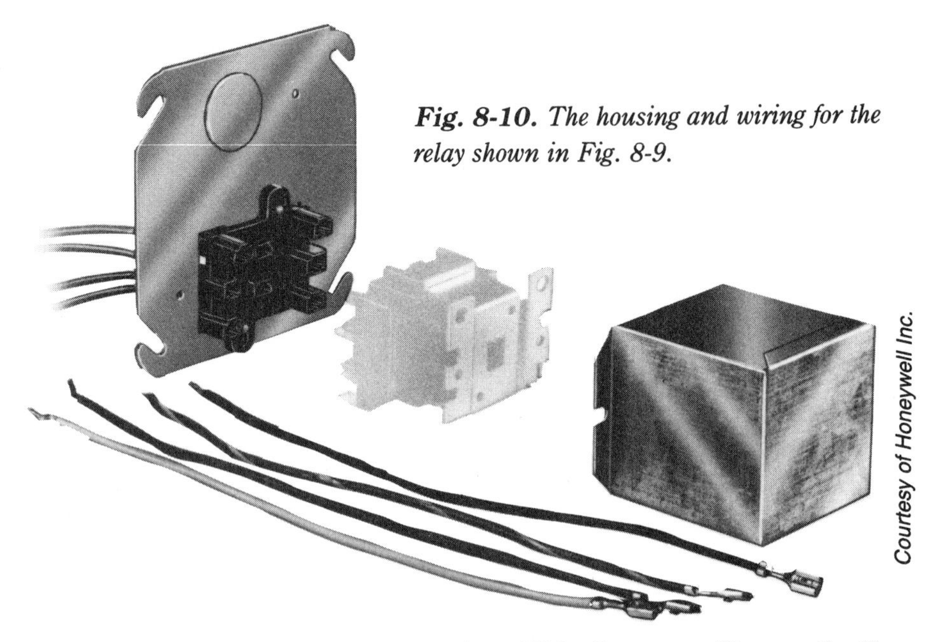

Fig. 8-10. The housing and wiring for the relay shown in Fig. 8-9.

Courtesy of Honeywell Inc.

unnecessary. Reliable thermostats, such as White-Rogers or Honeywell will give good service. Don't buy gadget types with numerous buttons that are difficult to program, are expensive, and do not always work the way they were programmed.

Wiring the thermostat in the furnace/air conditioner circuit is not difficult, but you should study the wiring diagrams closely (see Figs. 8-11 and 8-12). You will need diagrams for the furnace and the air conditioner because they must be correlated with each other. Contact the furnace manufacturer for a diagram, or your local furnace dealer might be able to furnish one.

TESTING THE NEW SYSTEM

Make sure that the refrigerant connections are tight, carefully check at the connections of the tubing and the cooling coil and condensing unit for leaks. Refrigeration and heating suppliers have a leak detection fluid, similar to the bubble liquid children use. The detector liquid cap has a dauber that is used to "paint" around any suspected leakage areas. Bubble liquid can be used with a paint brush in the same manner. This is a preliminary check that only locates larger leaks, other leak detectors are the "halide torch" and the electronic leak detector. The halide torch uses a propane tank for fuel and has a probe tube of rubber for searching out leaks. The head contains a flame that heats a copper ring in the flame. If the probe tube draws in any refrigerant, the light blue flame turns green, indicating a refrigerant leak. A large leak will turn the flame purple.

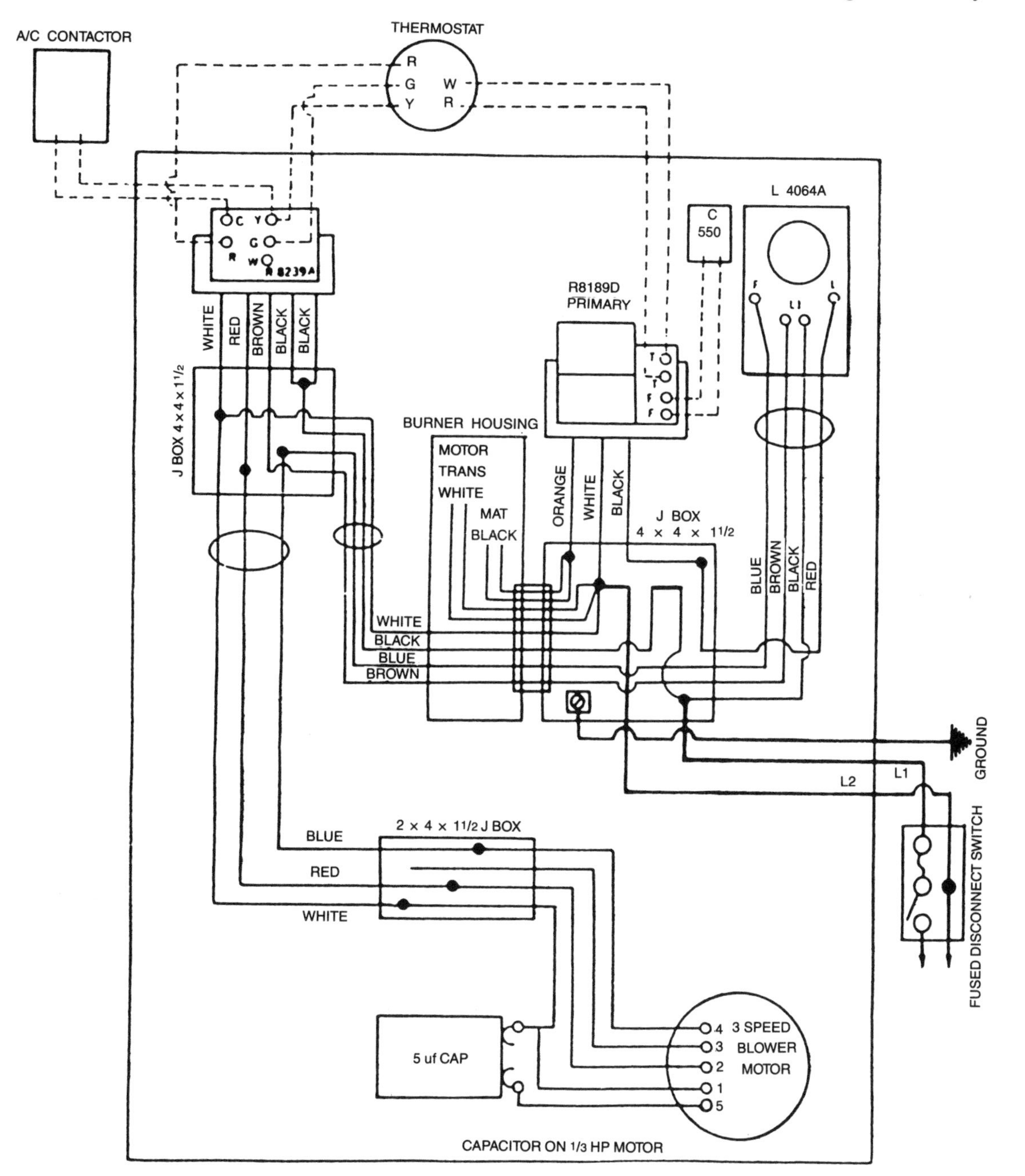

Fig. 8-11. *A wiring diagram for an oil-burning furnace with air conditioning.* Courtesy of Blue Ray Systems.

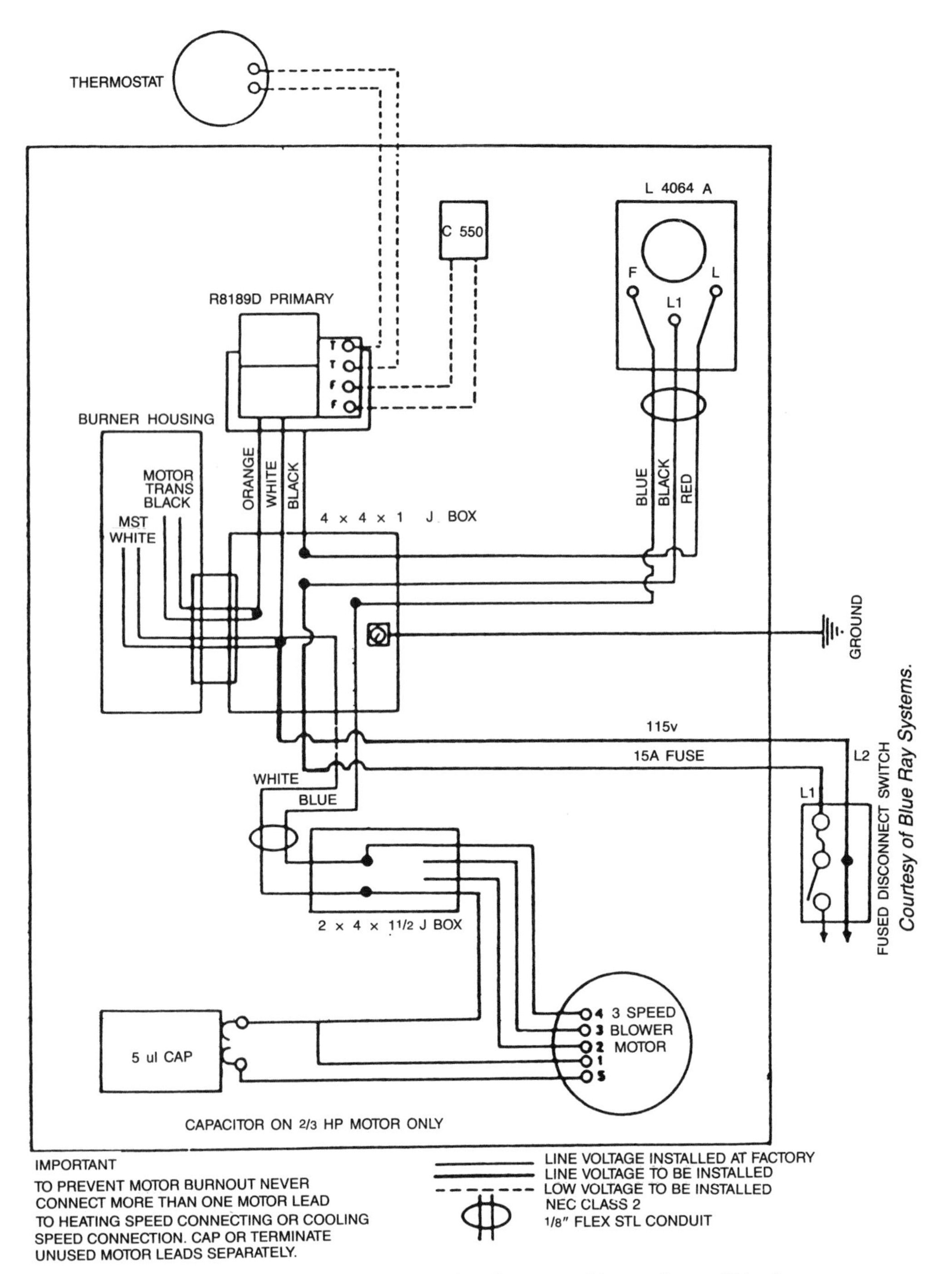

Courtesy of Blue Ray Systems.

Fig. 8-12. *A wiring diagram of an oil-burning furnace without air conditioning.*

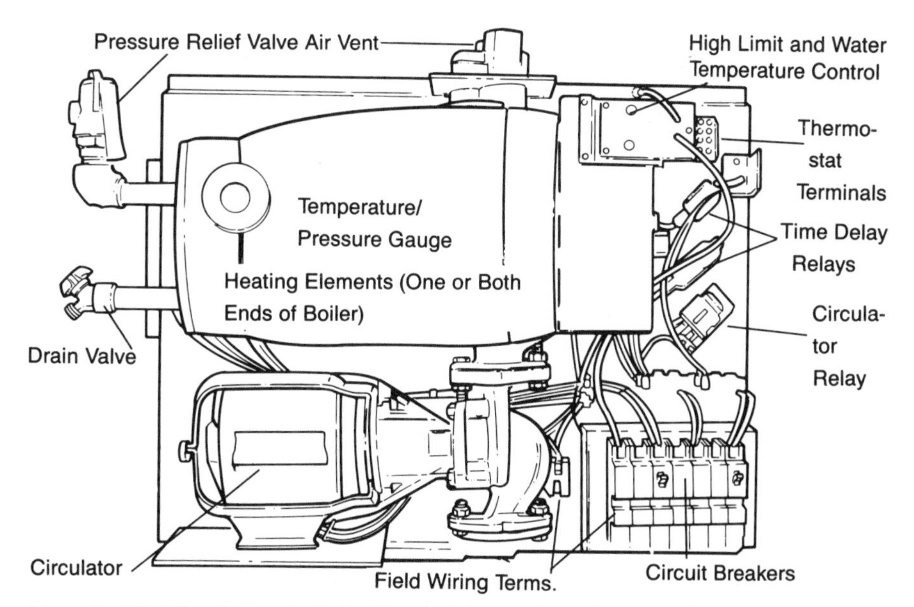

***Fig. 8-13.** This boiler is fueled by electric heating elements. It comes completely wired and piped. Only electricity and plumbing need to be connected.* *Courtesy of Electric Furnace-Man.*

The electronic leak detector also draws in air and refrigerant. Electronic components detect any refrigerant and give an audible and visual signal if a leak is found. These detectors are able to detect extremely small leaks, as little as a half ounce per year. These detectors cost as much as $175. The halide detectors cost about $30. Many years ago, the halide leak detectors were the only method available except for the leak detector fluid.

If you are unsure about turning on the air conditioner for the first time, you might want to call an air conditioning service company to do the start-up and check-out. A service person can do this for you at a cost of a service call. The installation manual furnished with the air conditioner will also give detailed instructions for the start-up and check-out. Read these instructions carefully, then decide if you want to do the work.

INSTALLING CENTRAL AIR CONDITIONING TO A FORCED, WARM-AIR FURNACE

It is comparatively easy to add central air conditioning to a forced, warm-air furnace. However, you will want to consider certain things when making this choice:

1) The condition and age of the present furnace.

2) The size and capacity of the furnace, and the attached ductwork.
3) The amount of space available above the furnace cabinet to install the cooling coil in the plenum above the cabinet, and if the plenum will need to be enlarged to accommodate the cooling coil.
4) Is there a floor drain near the furnace, or will condensate need to be pumped up and along the ceiling to a distant floor drain or sink? If this is so, an electric condensate pump will be needed.
5) Does the service entrance panel have enough extra capacity to handle the added load imposed by the air conditioning compressor?
6) Is the house well insulated? The ceiling insulation should be at least six inches, and preferably twelve inches, thick.

You need to consider all of these conditions before purchasing a central air conditioning system and installing it yourself. One of the first steps is to inquire at companies that sell heating and cooling systems for the do-it-yourselfer. Sears sells such systems as well as offering help in selecting one. Get all of the literature from the seller that you can and read it thoroughly before returning to the store. It can help you decide on the size and type of system best suited to your needs. Most medium-sized houses, up to 1,600 square feet, need a two and a half ton size unit, or 30,000 BTU capacity. (One ton of refrigeration equals 12,000 BTU.) This assumes adequate insulation in the walls ($3^1/_2$ inches) and ceiling (6 to 12 inches), as well as double-glazed windows or storm windows.

INSTALLING WINDOW AIR CONDITIONERS

A reasonable, low cost choice for air conditioning is the window air conditioner. These units offer relief from summer heat, at least for part of the house. Room air conditioners are frequently installed in bedrooms, so that people can sleep in comfort. You could add one or more units to be used individually as needed, rather than operating a whole-house air conditioning system.

Installing a window air conditioner is not difficult. First measure the window to determine if the window is large enough to accept the unit that you have selected. Most medium-size units fit standard windows. The lower sash must raise high enough to accommodate the height of the unit cabinet. The horizontal sliding window also must slide horizontally enough to accommodate the side-to-side unit dimensions of the unit. Brackets and side wings are furnished. Most installations present a neat look, but do not enhance the appearance of a living room.

The recommended exposures are north and east because south and west installations must operate in full sun, reducing their efficiency. If the unit is shaded in the afternoon and evening this will help, even though the unit is on the west or south side. Installation instructions explain how to support the unit and the amount of slant to the outside. The slant is required so that condensate will drain from the cabinet. Small, adjustable wings fill the space between the unit and the window frame. These wings must be sealed tightly with caulking compound so that they are rainproof. When the unit is installed in a double-hung window, the space between the two panes must be filled with a special filler material supplied with the unit. Caulking compound might be the only material that must be purchased locally. If the cord attached to the unit is too short to reach an electrical outlet, buy an extension cord specified for use with an air conditioner. Do not use a light duty cord, because you will damage the unit. Use #14-3 cord for 15 amps.

Window air conditioners vary in price from $99 to $500, depending on the capacity and extra features. The smallest size, one half ton, or 6,000 BTU, might be sold as a "carry home plug-in" model. The largest size, a two ton, or 24,000 BTU, is not a carry home model and must be plugged into a special circuit for this unit only, protected for 20 A at 240 V and wired with 12-2, with a ground cable. In any case, a time delay fuse or a circuit breaker rated at 20 A must be used. An air conditioning unit drawing 8 A will, on starting, draw momentarily from three to five times its rated amperage (24 A to 40 A). This current draw lasts for only one or two seconds, until the motor is up to speed. This much higher amperage might be shown on the unit nameplate as "Locked Rotor Amps." A time delay fuse or a circuit breaker is able to handle this current inrush, and should not blow the fuse or trip the circuit breaker. Check your wiring to determine what size wire is feeding the outlet to be used for the air conditioner.

I once installed a window air conditioner in an older house. While doing the installation, I noticed an outlet below the window that could be used for the unit. After the installation was completed, I decided to go to the basement and check on the wiring feeding the outlet. To my surprise, the outlet was wired with lamp cord that was plugged into a basement ceiling light. Using that outlet could have caused a fire in the wall or damaged the compressor. I informed the owners that I would not plug the unit into this outlet, as it would be dangerous. The owners were very upset, even though I referred them to a reliable licensed electrical company to correct the problem. Beware of operating a window air conditioner on an overloaded or undersized circuit, damage can result. Reduced unit life or a burned out compressor motor is usually the result of operating a unit on low voltage. Most electrical equipment operates satisfactorily on 120V or 240V, plus or minus 5 percent voltage. Any more or less could cause damage. The wiring to an electrical outlet that is undersized for the load and at the same time is fused oversize, or the circuit breaker is oversize, is liable to cause a fire inside the wall or in the attic.

If you already have forced, warm-air heating, and are planning to install more than three window air conditioning units, check the cost of a complete house air conditioning system. If you install central air conditioning yourself, the cost of three window units could possibly equal the cost of central air conditioning. Spring and fall are ideal times to install central air conditioning. The weather is mild, there is no need to hurry, and there is less mess tracking in and out at these times.

A satisfactory arrangement is to install a large, 18,000 BTU window unit in an attic gable end, and run a duct from the cool air discharge opening of the unit to an attractive grille in the ceiling of the living room. A return air grille is then installed in the ceiling of a rear hallway. The joist space is used as a return duct back to the unit, then connected to the return air opening of the unit.

This set-up operates very well and provides satisfactory cooling for the first floor bedrooms, living room, kitchen, and other areas. A 3/4 ton, 9,000 BTU unit can be installed in the second floor window, which would keep the bedroom very comfortable during hot weather.

Sometimes, it is more convenient to install a window air conditioner through an outside wall. This eliminates blocking a window and can provide a better appearance for the whole room. This involves more work, however, because the opening for the unit has to be cut in the wall and framed with 2 × 4 studs. After removing drywall or lath and plaster, one vertical stud must be cut at the general location. Removing this stud leaves a horizontal width of about 30 1/2 inches, wide enough for a medium-size air conditioning unit. If you use a larger unit, you might have to remove two studs to provide for the greater width. It is very important to take careful measurements before doing any cutting.

The unit must be slanted slightly to the outside so that condensate will drain away from the unit, therefore the opening must allow for this tilt of the unit itself. Start making the opening by removing siding or bricks from the outside. Then remove enough sheathing so that the center 2 × 4 can be cut short enough to allow the "cripple" cross studs to be set in place. The center stud or studs must be cut one and a half inches above and below the finished opening to conceal all of the framing behind the drywall or other wall finish. By removing extra siding and sheathing from the outside, the center stud or studs can be cut with a power saw. The cripples should be fastened in place using angle brackets or toe nailing. Angle brackets will be easier. Nail the cripples to the cut ends of the center stud or studs with 16d nails. A filler stud or a thinner vertical filler piece might be needed on one or both sides of the opening to provide a snug fit of the unit casing. See Fig. 8-14.

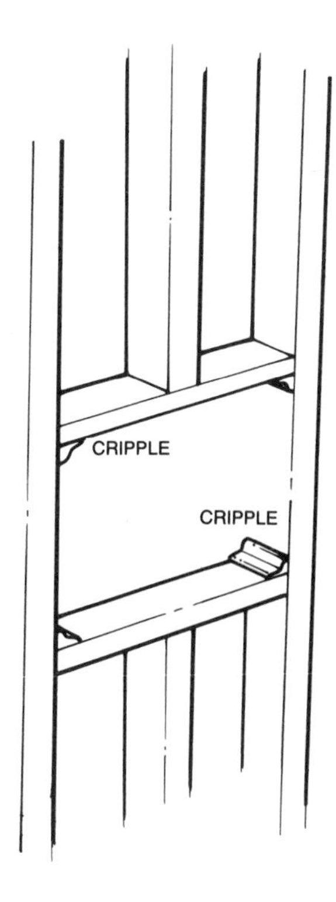

Fig. 8-14. *The framing arrangement for installing a window air conditioner through a wall.*

The finished opening must be a snug fit, with drywall or lath and plaster extending to the sides of the unit when installed. The completed opening should be trimmed with molding, for a neat appearance. Replace the sheathing and siding or bricks on the outside of the opening. Some manufacturers can furnish sleeves made for through-the-wall units as a permanent installation, which you can caulk and make weatherproof.

9

Heat Pumps

FOR SOME TIME NOW, HEAT PUMPS HAVE BEEN ADVERTISED AS THE ANSWER TO heating homes and commercial buildings. Utilities and heating contractors claim that operating expenses compare with other forms of fuel for heating. Claims are made that the heat pump can provide summer cooling. Actually, the heat pump is not efficient at either heating or cooling. The heat pump is a compromise between straight air conditioning and a fossil-fueled forced air furnace. It is true that heat pumps do operate satisfactorily in mild climates where temperatures do not drop below 35°F. At lower outdoor temperatures, it is necessary to provide auxiliary heat. Electric resistance heat is available. This is *extremely expensive* to use. If a gas- or oil-fired furnace is used, money will be saved by using this system as the primary heat source rather than buying a heat pump, then a straight summer air conditioner can be installed. This type of heating and air conditioning is the most economical.

Extensive research has been done on the projected life of a heat pump. It has been determined that the average life of a heat pump is 8 to 10 years. As the servicing and repairing of it requires a specially trained technician, this becomes very costly. The system operates, during the heating season, at very extreme conditions. Below 35°F the outside cooling coil will frost up. To overcome this situation, the system goes into a ''defrost'' cycle to melt the frost from the cooling coil so that it can again extract heat from the cold air. Also at this time, the system turns on auxiliary heat, either electric resistance heating, oil, or gas heat. If there *is* a fossil fuel (gas or oil) forced air furnace to provide auxiliary heat, then the heat pump should never have been sold for this house. Gas or oil fired furnaces have an average life of *20* years, twice that of a heat pump. A once-yearly checkup and cleaning will provide trouble-free heat all winter. Many gas furnaces have never required service in their lifetime.

Although the general principles of operation are standard, there are so many makes and variations in design of heat pump components that I decided to give only a general overview of heat pumps. In effect, the heat pump is the exact reverse of an air conditioning system. An air conditioning system absorbs heat from an area or building. A cooling coil is installed in a forced, air-furnace plenum (the ductwork just above the top of the furnace). This coil is at 40°F when the air conditioning is operating. Heat travels from the warmer air to the colder coil, thus the cold coil absorbs the heat of the air. If 80°F air passes over a cooling coil at 40°F, the coil will absorb the heat of the air; the air leaving will be at about 55°F. This air is then delivered to the area being cooled.

The heat pump works the reverse of the air conditioning system. The cooling coil of the heat pump is the *outside* coil. During the cold months, this cooling coil absorbs heat from the cold, outdoor air. This might seem impossible, but the outdoor coil *is* colder than the outdoor air and absorbs the heat from the outdoor air. This system is feasible only if the outdoor air is above 35°F. If the outdoor air is below 35°F, additional heat must be provided. Although there is heat available in the cold, outdoor air below 35°F, there is not enough to allow the heat pump to operate.

SUPPLEMENTARY HEAT

The heat pump cannot provide heat if the outdoor air is below 35°F. Supplementary heat then becomes necessary to keep the heated area at 68°F. Two forms of supplementary heat are available: Electric resistance heating coils, or oil- or gas-fired, forced, warm-air furnace. Electric resistance heating is extremely expensive, and if oil or gas is available, these types should be used. If you now have an operating gas or oil furnace, you should continue to use this furnace. Modern gas- *and* oil-fired furnaces now have an efficiency of 90 percent or greater. These furnaces are more economical in the long run.

HEAT PUMP OPERATION

The typical heat pump is generally sold as a combination heating/cooling system. In the cooling mode, the heat pump consists of an evaporator (cooling coil) in the furnace plenum. This coil is connected to the remote, outdoors condensing unit and consists of a compressor and condensing coil (outdoor coil). As the unit operates, the compressor draws gaseous refrigerant from the inside cooling coil and compresses it to a higher pressure, 210 pounds per square inch (PSI). This gas goes to the condensing coil outdoors and is condensed to a liquid, giving up its heat to the air blown over the coil by the fan. From here, the liquid is pumped to the indoor evaporator (now the cooling coil) in the furnace.

Before entering the cooling coil, the liquid passes through a filter-drier, various valves, and a thermostatic expansion valve (TEV) before entering the cooling coil. The TEV reduces the liquid refrigerant pressure to about 69 PSI. At this point, it changes to a gas, absorbing heat from the air passing over the cooling. This air is cooled and distributed to the area being cooled. This is the cooling mode. See Figs. 5-16 and 5-17.

When heating is required, the operation of the system is reversed. The indoor evaporator (cooling coil) becomes the heating (condensing) coil. At the same time, the outdoor condensing coil becomes the evaporator (cooling coil) and the flow of the refrigerant is reversed. All of this is accomplished by a reversing valve or valves located in the cabinet housing the condensing unit and coil. The reversing valves exchange the roles of the two coils. The indoor coil now heats the area inside, while at the same time, the outdoor coil absorbs heat.

Many heat pump/air conditioning systems are operated manually; some are completely automatic. The manual system is the better. The switchover is accomplished by switches on the thermostat subbase. These are: 1) Heating/Cooling. 2) Fan, Auto./Off/On. Thermostats have one scale and one or two pointer levers, one for heating and one for cooling. You can then select which setting you need depending on the need for heating or cooling.

TYPES OF HEAT PUMPS

Heat pumps are usually installed as "air-to-air" types. There are also heat pumps that extract heat from buried pipes that are filled with circulating water. This type of heat pump is a "water-to-air" system, as the water gives up its heat to the air in the building. This type of system is better than an air-to-air system because more heat is able to be extracted from the warmer water. Unfortunately, this type of system is very expensive because the pipe must be buried below the frostline and connected to the heat pump system.

A HEAT PUMP'S LIFE CYCLE

Over the years the cost of (heat pump) ownership is more than just the initial purchase and the monthly energy bills. Maintenance and replacement costs should also be included. In order to evaluate the total cost of ownership it is helpful to consider all the costs over the life of the appliance. This is referred to as the "life cycle cost." When comparing alternatives that have different lives, a fixed study period can be used. Table 8-1 provides an example of a life cycle cost comparison over a 15-year study period. The ownership costs of a gas system are compared with an electric heat pump system over a 15-year period using constant dollars (Constant dollars assumes a constant value of the dollar. This is reasonable as long as all the costs are summed and brought into the current year. See Table 8-1).

SERVICING HEAT PUMPS

Except for the refrigeration cycle, you should be able to service your air conditioning/heat pump unit. The refrigeration cycle requires extensive experience and training, and an amateur could easily damage the motor/compressor. Specialized and expensive testing equipment is also required. If the trouble is in the refrigeration cycle, I recommend you employ a licensed refrigeration/air conditioning service technician to do the work.

The outdoor (condensing) unit does not require any service except oiling of the condenser fan. Check any couplings and flarenut connections for any signs of leakage, indicated by oil leakage (this also indicates leakage of refrigerant). Use a soap solution or child's bubble liquid brushed on the suspected leaking areas. If a bubble forms that indicates a leak at that point. Call a refrigeration technician to make a thorough leak check of the system.

A heat pump operates similar to air conditioning, except there are additional controls and valves. This makes the system much more complex than straight air conditioning. During winter, the heat pump system reverses its function from that in summer. The cooling coil in the furnace becomes the ''condenser,'' and gives off heat and the outdoor condenser becomes the cooling coil (actually a heat absorbing coil). Figures 5-16 and 5-17 illustrate the changeover from summer to winter operation and back from winter to summer again.

10

Space Heating Appliances

SMALLER HOMES WITHOUT CENTRAL HEATING, COTTAGES, AND OTHER SMALL dwellings generally use a form of space heater. These take the form of freestanding heaters, fired by natural gas, liquified petroleum gas, fuel oil, electricity, and firewood. Most of these heaters stand in the main room, at least three feet (away from walls), thus ''freestanding.'' Wood burning stoves can take the form of a freestanding stove, or can be installed in an existing fireplace and use the fireplace chimney. A large percentage are automatic in operation and only require minimum care. **Caution**: All of these units, except the electric heaters, **must** be vented, using a flue pipe connected to a chimney or directly to the outside. Unvented fuel-burning heaters are extremely dangerous devices and can cause death from carbon monoxide accumulation!

SPACE HEATERS

The modern space heater is a designed and tested appliance, with safeguards to prevent tipping, overheating, and producing poisonous carbon monoxide (CO). Space heaters can be used with kerosene, natural gas, and Liquified Petroleum Gas (LPG), commonly called bottled gas, depending on the make. **Caution**: It is strongly recommended that **no** space heater of any type be operated in an enclosed area *unless* it is properly vented to a masonry chimney or prefabricated chimney. An *unvented* space heater in a closed space, even though properly maintained and operated, will surely produce carbon monoxide (CO), a deadly, colorless, odorless gas, also produced by automobiles. Small space heaters, used in campers and other restricted, closed spaces have produced this poisonous gas, causing many deaths.

Heaters are still sold that cannot be vented. **Do not buy** these heaters. The only space heater that does not need to be vented is the electrical resistance heater with the glowing coil. These heaters are very safe, and will shut off if tipped over. Oil-filled baseboard radiation or portable, radiator-shaped space heaters can also be used. These heaters have a permanently sealed oil reservoir and an electrical resistance heating element. These heaters become comfortably warm to the touch, but will not cause burns when touched. This makes them safer around small children than the glow coil type. Many of these heaters are designed to be permanently wired, and need a 208V to 240V electric supply. The portable type is also made and is designed to be plugged into a 120V outlet. Figures 10-1 and 10-2 show these portable heaters.

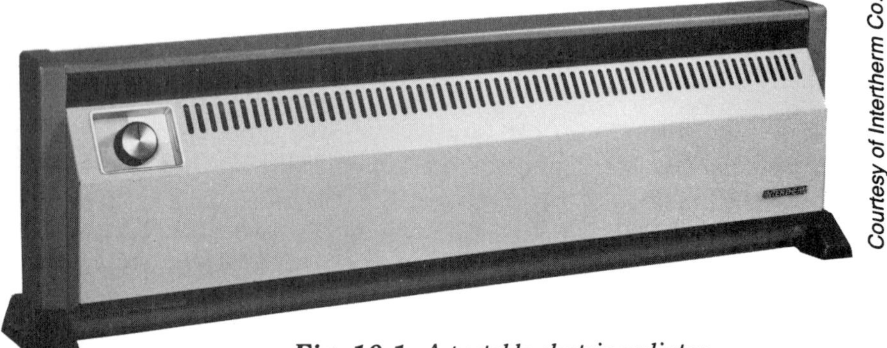

Courtesy of Intertherm Co.

Fig. 10-1. *A portable electric radiator.*

Oil-Fired, Constant-Level Valve Space Heaters

Oil-burning space heaters with a constant-level valve are still in use in many parts of the country. These heaters are available in two styles, the radiant type and the circulating type. The radiant type has a firebox connected to a flue pipe, the constant-level valve, and an oil reservoir. There is no casing around the firebox, and it would be dangerous if there are young children in the house. A better style has a casing around the firebox, which does not get hot to the touch. This style circulates air through the space separating the firebox and the casing by gravity. This heats the room much better than the radiant type. These heaters are satisfactory for cottages and homes that have only one floor. Maximum capacities are 50,000 BTU per hour. The type of fuel oil used is No. 1. This is equivalent to kerosene. The type of fuel oil used with the power-operated oil burners is No. 2, and *must not* be used in these heaters.

Courtesy of Intertherm Co.

Fig. 10-2. *A portable water-filled electric radiator.*

The design of the heater is: 1) A fire pot, usually ten inches in diameter. A flange sits on top of the fire pot. This reduces the size of the top opening and contains the flame in the fire pot. 2) An oil storage tank, mounted on the rear of the unit. 3) The constant-level control valve. This valve, erroneously called a "carburetor," is fed by gravity from the storage tank. The valve operates by maintaining a pool of oil in the bottom of the fire pot. The valve has a float valve in its reservoir, the bottom of which is at the same level as the bottom of the fire pot, thus the level of the oil in the fire pot is maintained at the same level as that in the valve body. The valve has an adjustable knob on the top for raising or lowering the level of the oil in the fire pot, according to the needs of the house. The oil level in the fire pot varies from $1/8$ inch to $3/16$ inch. The cover of the control has numbers from one to six, for adjustment of the height of the flame, by means of this knob. This control is manually operated, and constant monitoring is necessary. A number one or two setting is recommended when the house is not occupied. It is sometimes convenient to connect an outside, 220-gallon oil tank to the constant level valve. Use $3/8$ inch OD copper tubing. Include a special "fire valve" in the line just before it enters the constant level valve. In case of fire or excess heat, the valve will snap shut, preventing oil from feeding any fire.

You might not be able to purchase these heaters any more, but it is well to know how to service them properly if you have one. The fire pot should be cleaned once a year after the heating season. Usually, the line to the fire pot can be disconnected and the pot lifted and carried outdoors for cleaning. The flue pipe should be inspected and replaced, if necessary. There will be a draft stabilizer installed in the flue pipe immediately after it leaves the heater. This consists of a flue pipe tee, the same diameter as the flue pipe. The side of the tee has a swinging damper with an adjustable weight. This weight can be adjusted to give more or less draft over the fire. Too little draft will cause a smoky fire, too much draft can blow out the fire. The door of the heater usually has a window so you can observe the flame. Carefully crack this door open, the handle might be hot. Hold a lighted match or candle near the crack. The flame should be drawn in toward the fire. This indicates a good draft. If the flame is not drawn in, or is blown back out, the draft is not sufficient. Adjust the weight slightly on the draft stabilizer, so that the damper is closed slightly more than before. Check the draft again with the flame.

Keep in mind that water in fuel is not a hazard, it just stops the flow of oil in this type of heater. It is important to keep an outdoor oil tank as full as possible to prevent condensation of water in the fuel tank.

WALL FURNACES AND FLOOR FURNACES

Wall and floor furnaces are gas- and oil-fired furnaces that are usually found in smaller homes and cottages. Often, they are installed in an added room on a larger house. These furnaces must be vented, as required by all building codes. A cement composition flue pipe in an oval shape was used extensively in previous years. The new, metal double-walled insulated oval flue pipe has replaced the older type because of the ease of handling. The oval shape of both types of flue pipe allows it to be run inside the hollow walls, and be concealed. See Figs. 10-3 and 10-4.

Either type of furnace can be controlled by a wall thermostat, making them fully automatic. Wall type furnaces are designed to fit between studs in a wall. Some types need to have a wall space framed in, as they are wider than the standard stud spacing of 16 inches on center. The gas line can be brought to the wall space from below, either a crawl space or a basement. A flexible connection line approved for use on gas appliances can be used to make the final connection. The gas line from the supply or meter might be 3/4 inch or 1/2 inch black iron pipe. Instructions furnished with the furnace make recommendations for running the flue pipe and gas line. Run black iron pipe close to the furnace location, and install a shutoff valve on this line. The flexible connector line from this valve to the furnace connection can now be installed. You may need adapter fittings to connect the shutoff valve, and to the furnace valve. These fittings are available where you purchased the flexible connector line. See Figs. 10-1 and 10-2.

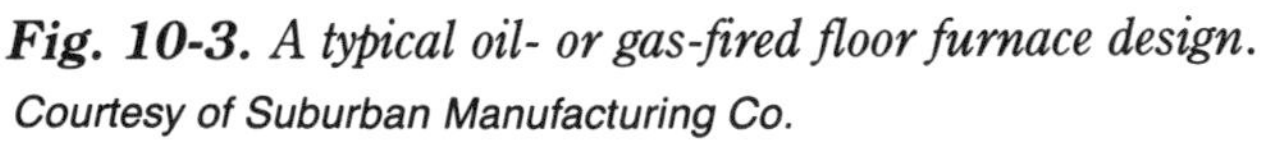

Fig. 10-3. *A typical oil- or gas-fired floor furnace design.*
Courtesy of Suburban Manufacturing Co.

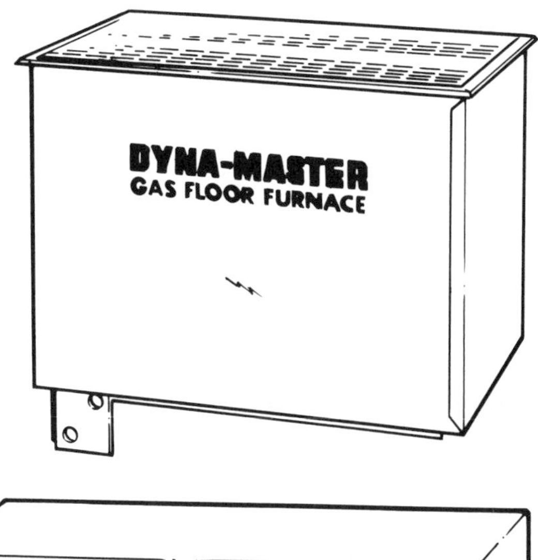

Courtesy of Perfaction Products Co.

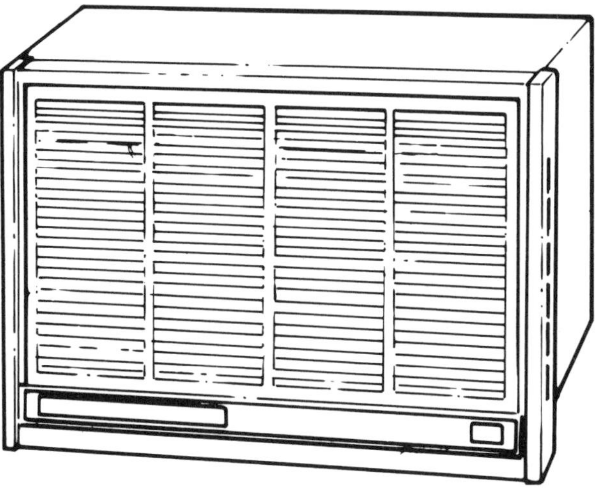

Fig. 10-4. *Two, through-the-wall models. The one above is a gas heater with a sealed combustion chamber. It takes combustion air from the outdoors and then later returns it* Courtesy of Suburban Manufacturing Co. *The other model is a typical, vented wall furnace that uses a standard flue pipe through the roof because the furnace is in the inside wall.*

The floor furnace is similar to the wall furnace except that it is suspended between floor joists, usually in an open area, such as a living room or hallway (see Fig. 10-5). The furnace has a floor grille similar to a floor register, and can be walked upon. The design of the floor furnace is somewhat different from the wall furnace, such as the flue connection. Although the installation directions are specific regarding flue pipe connections, I recommend that you install a special cap on the top of the flue. The flue must be two feet or more above the peak of the roof. This also applies to the flue from a wall furnace. The gas line to the floor furnace can be piped entirely with black iron pipe. Remember, the proper sequence of fittings for piping gas lines is: Gas shutoff valve, nipple, union, nipple, tee (with side opening facing down and having a nipple and cap screwed into the side opening), a section of pipe, then the automatic valve of the furnace. Figure 10-3 shows a floor furnace.

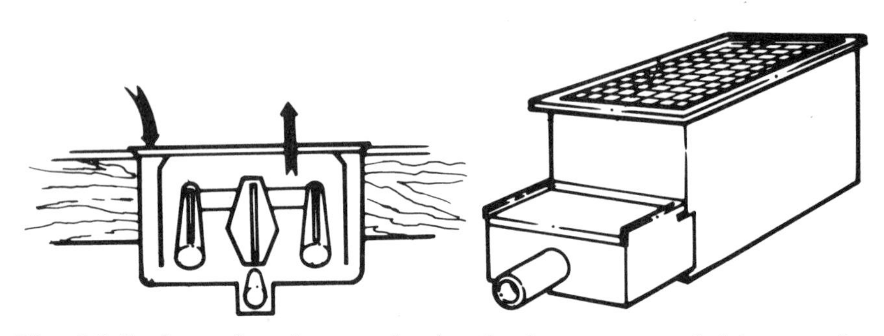

Fig. 10-5. *A gas floor furnace showing the furnace suspended between floor joists.*

The furnace has an automatic valve, a manifold, and two or more burners, plus the pilot burner. A nameplate giving lighting directions can be found on the front of all furnaces. Follow the recommended procedures for cutting and threading iron pipe, applying pipe compound or Teflon tape to the threads, and joining the pipe and fittings together correctly. Do not allow the gas line to put any strain on the automatic valve body. The pipe must be supported independently on its own hangers.

The oil-fired floor or wall furnace operates similarly and is installed in the same manner as the gas-fired models. The type of fuel used is the main difference. The oil model will have a constant level oil valve, usually operated manually. The floor model will have a long rod extending from the knob of the constant level valve to the floor grille, in a recessed corner for controlling the flame size. The wall model's constant level valve is accessible through a small door in the grille.

Note that a gas-fired wall furnace is made as a direct vent type (see Fig. 10-6). This furnace takes combustion air from the outdoors and vents the products of combustion to the outdoors. Consequently, no flue pipe is needed, and the unit is complete. All that needs to be done is to pipe the gas to the unit. The thermostat is integral with the furnace. These units have been manufactured for many years and have proved very satisfactory for single added rooms and other supplemental heat. Figure 10-4-A illustrates a typical through-the-wall heater. This heater has a sealed combustion chamber and uses no room air. Figure 10-5 shows a floor furnace installation. Figure 10-6 shows a through-the-wall side view installation of a wall furnace.

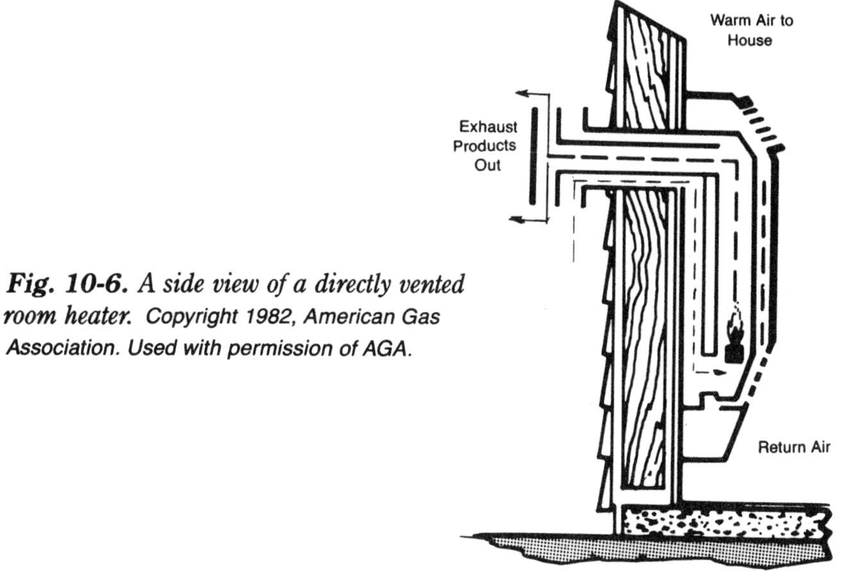

Fig. 10-6. *A side view of a directly vented room heater. Copyright 1982, American Gas Association. Used with permission of AGA.*

WOOD BURNING STOVES

Wood burning stoves have become very popular both for remote and urban areas. While comparatively expensive to buy, these stoves can save you money if a supply of wood is available on the property. Areas near small towns and suburban areas usually have wood for sale at reasonable prices. Be sure the wood to be purchased has been seasoned for at least six months under cover, or you could season it yourself if you buy it six months in advance of use.

The use of a wood stove for space heating requires more care and attention than either gas- or oil-fired automatic heating equipment. Wood stoves are much more efficient if a catalytic combustor is installed in the stove. This device assures a more complete combustion, providing a hotter fire for the same amount of wood. In addition, 90 percent less creosote is deposited in the chimney or flue. Oregon and Colorado have enacted strict antipollution laws which have been in effect since 1986 and 1987 for catalytic combustors. New standards are in effect as of July 1, 1988 for both states. Federal regulations go into effect July 1, 1990. The Environmental Protection Agency standard for stoves with catalytic combustors is a maximum allowable particulate emission of three grams per hour. For noncatalytic combustors, the maximum allowable particulate emissions is seven and a half grams per hour. I recommend that you install a catalytic combustor. You can purchase one for a cost about $100 to $175. The stove will operate more efficiently and cleanly, and more of the unburned gases are captured and burned, increasing the efficiency by 25 percent. Some stove manufacturers claim that using the combustor can pay back in one heating season.

Installing a Wood Burning Stove

Wood and coal burning stoves are installed differently than a furnace or boiler. Because these stoves will be installed in the living areas, certain safety and clearance rules *must* be observed. Chimneys are particularly hazardous if these rules are not followed. In older homes, the chimney might not have a clay tile liner. This chimney construction is extremely unsafe and can cause a fire in the attic or inside the house walls. Loose mortar can fall out from between the bricks at some point, leaving a direct opening in the chimney wall. Sparks from the wood fire might ignite wood building construction near the chimney and cause a bad fire. This same condition is also dangerous even when automatic equipment is used. See Fig. 10-7.

As mentioned before, there must be a clay tile liner, in perfect condition from top to bottom. Chimneys with defective or broken tile can be reused by installing a flexible, prefabricated metal liner or a sectional rigid metal liner. Both types of liners are double or triple wall construction and are listed by the Underwriters Laboratories (UL). No matter which type you choose, I recommend that insulation be packed around the liner between the liner and the flue space surrounding it. The metal chimney liners are listed as a ''Class A All Fuel Chimney'' by the UL. If a new masonry chimney is to be built, it should be built in the interior of the structure, not against the outside of the structure. Built properly, this chimney is better, as a good draft is developed, and the chimney will last a very long time. Again, the chimney must have the clay tile liner installed to be safe. Figure 10-8 shows sections of metal chimney and how they fit together. Figure 10-9 shows an existing chimney for a wood stove. Also see Figs. 10-10 through 10-12.

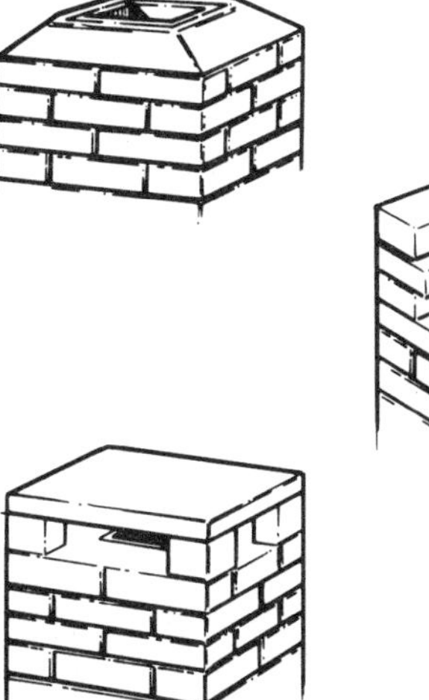

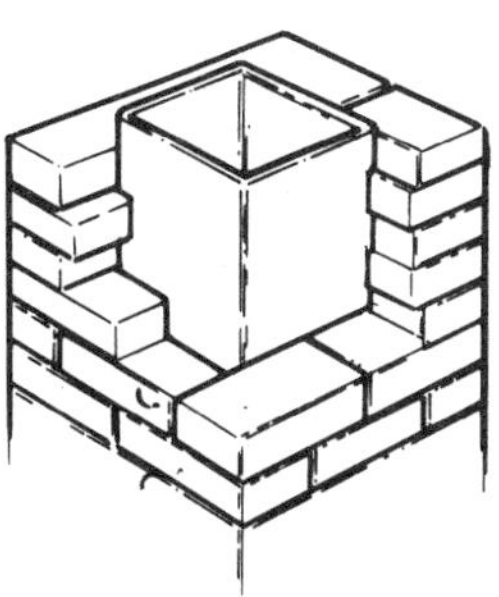

Fig. 10-7. The accepted design of a masonry chimney. The tile liner is shown on the right, and the correct way to finish a chimney top is shown on the left. To prevent debris from falling into the chimney, a cement slab, supported by four half-bricks, is built on to the top. Courtesy of Consolidated Dutchwest. Used by permission.

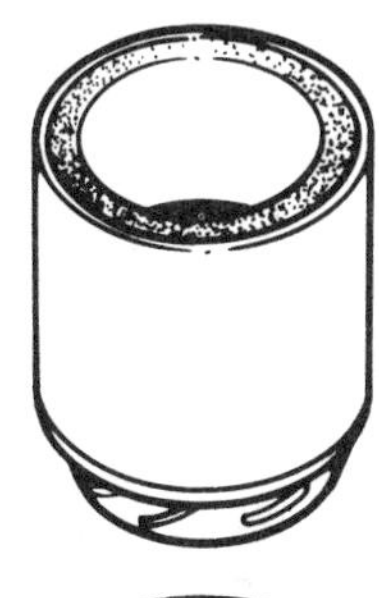

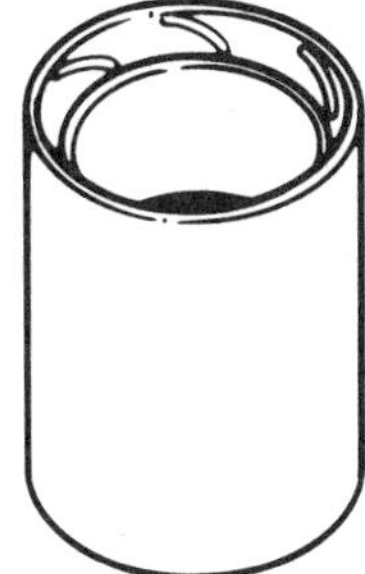

Fig. 10-8. An insulated double-wall chimney flue pipe. Sections are fitted together, making for a very smoke-tight installation. When used in a masonry chimney, the space between the new, round metal chimney, and the square masonry chimney, must be filled with insulation. Courtesy of Consolidated Dutchwest.

Chimney Safety

Burning wood in a properly designed wood stove, and using a catalytic combustor, some creosote will still deposit on the interior walls of the chimney. An annual cleaning of the chimney is therefore necessary and should be done by a professional chimney sweep. Currently, stoves and fireplaces are very popular, and chimney sweeps are listed in the classified phone book in most parts of the country. A chimney sweep inspects the chimney and can determine if it is safe

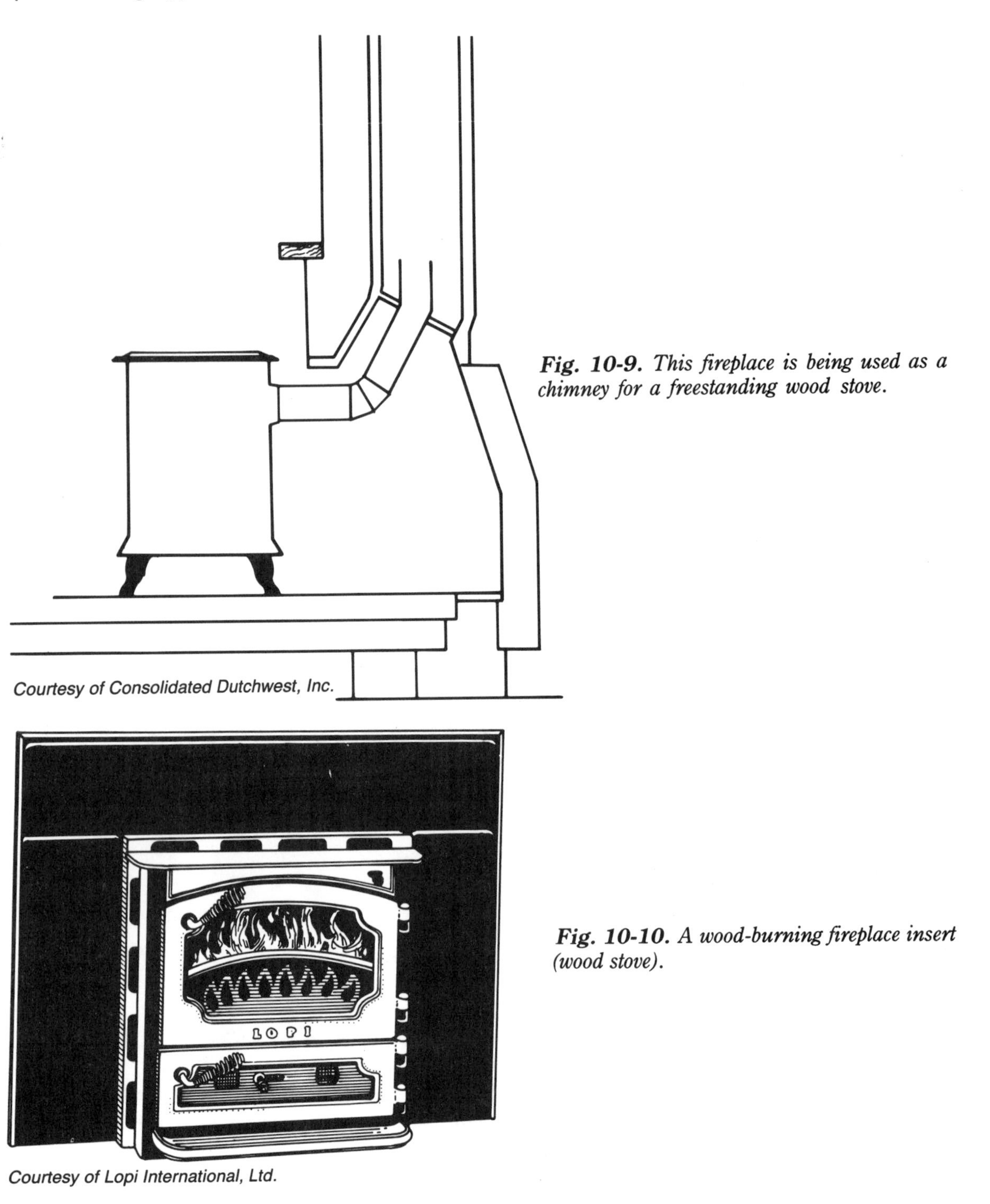

Fig. 10-9. This fireplace is being used as a chimney for a freestanding wood stove.

Courtesy of Consolidated Dutchwest, Inc.

Fig. 10-10. A wood-burning fireplace insert (wood stove).

Courtesy of Lopi International, Ltd.

Courtesy of Consolidated Dutchwest, Inc.

Fig. 10-11. *A freestanding wood stove.*

to use, or if a prefabricated liner is necessary. Proper and adequate clearance must be maintained between the wood stove and all combustible surfaces. A fire-proof hearth must be constructed for the stove to sit on, extending at least 16 inches beyond the stove on all sides. A suitable hearth pad can be fabricated, using $^{1}/_{4}$-inch mineral board (available at hardware stores and home centers) on $^{3}/_{4}$-inch plywood base, covered with 24 gauge sheet metal. The top should be covered with slate or quarry tile and a trim board attached around the four sides. Refer to Fig. 10-13 which shows the construction. Walls adjacent to the stove must be protected if they are less than 36 inches from the stove. If the walls are less than the 36 inches, a shield must be installed, spaced away from the wall at least one inch, using screws and pipe or tubing spacers. The shield must be made of a noncombustible material such as sheet metal or mineral board. Figure 10-13 illustrates the hearth construction.

Operating a Wood Stove

Many factors influence the efficiency of a wood stove:

1) Stove design.
2) Proper operation.
3) The kind of wood used and its condition.

4) Adequate air for combustion. Most stove manufacturers make an excellent product with airtight construction. As mentioned, adequate air for combustion is very necessary, even with oil- and gas-fired equipment. A very tightly built house will starve any fuel burning equipment of oxygen. Wood and coal burning stoves produce carbon monoxide (CO), an odorless, deadly gas. The presence of CO is **EXTREMELY DANGEROUS** inside a closed structure. An opening *must* be provided to the outdoors for fresh air to enter the furnace area. For each 4000 BTU per hour, one square inch of *free* area *must* be provided. Free area is defined as the total area of the openings in the grille or other material. Metal louvers and grilles are assumed to have 60 percent free area. See Figs. 10-10 through 10-15.

The best fuel for the stove is medium-size seasoned hardwood, cut and stored under cover for at least six months. Buy from an established supplier to obtain dry wood (15 percent to 20 percent moisture content).Some manufacturers make wood stoves that can be converted to burn coal. One manufacturer states that the conversion takes only minutes. This is advantageous in extreme cold weather, as a coal fire lasts twice as long as a wood fire, up to 20 hours. Anthracite (hard) coal is best, but high quality bituminous (soft) coal is nearly as good.

Recent developments in wood stove design by one company, LOPI, claim high efficiency and low emissions. This design incorporates a heat retentive masonry baffle within the firebox to promote secondary combustion. This secondary combustion burns volatile gases, releasing more heat and increasing the efficiency of the freestanding wood stove and the fireplace insert heater.

These wood stove/insert appliances meet the stringent emission standards of the state of Oregon without using the expensive catalytic combustor. The combustor needs to be replaced periodically at a cost of about $100. The masonry baffle does not need replacement. These heating units are manufactured by LOPI International, LTD, 10850 117th Place, N.E., Kirkland, WA 98033, Telephone 1-800-426-3915. Figure 10-6 shows LOPI fireplace inserts and fireplace vented stoves. Freestanding stoves are shown in Figs. 10-11 and 10-12. Figure 10-4 shows a side view of a wood stove using a fireplace chimney as a flue. Shown in Fig. 10-14 is a cut-away view of a wood stove. The catalytic combustor is shown at the top center as a perforated cylinder.

LOADING WOOD AND STARTING AND MAINTAINING A WOOD FIRE

Most Federal airtight stoves load wood from both the front and side. Stoves with front and side door loading are usually easier and more convenient

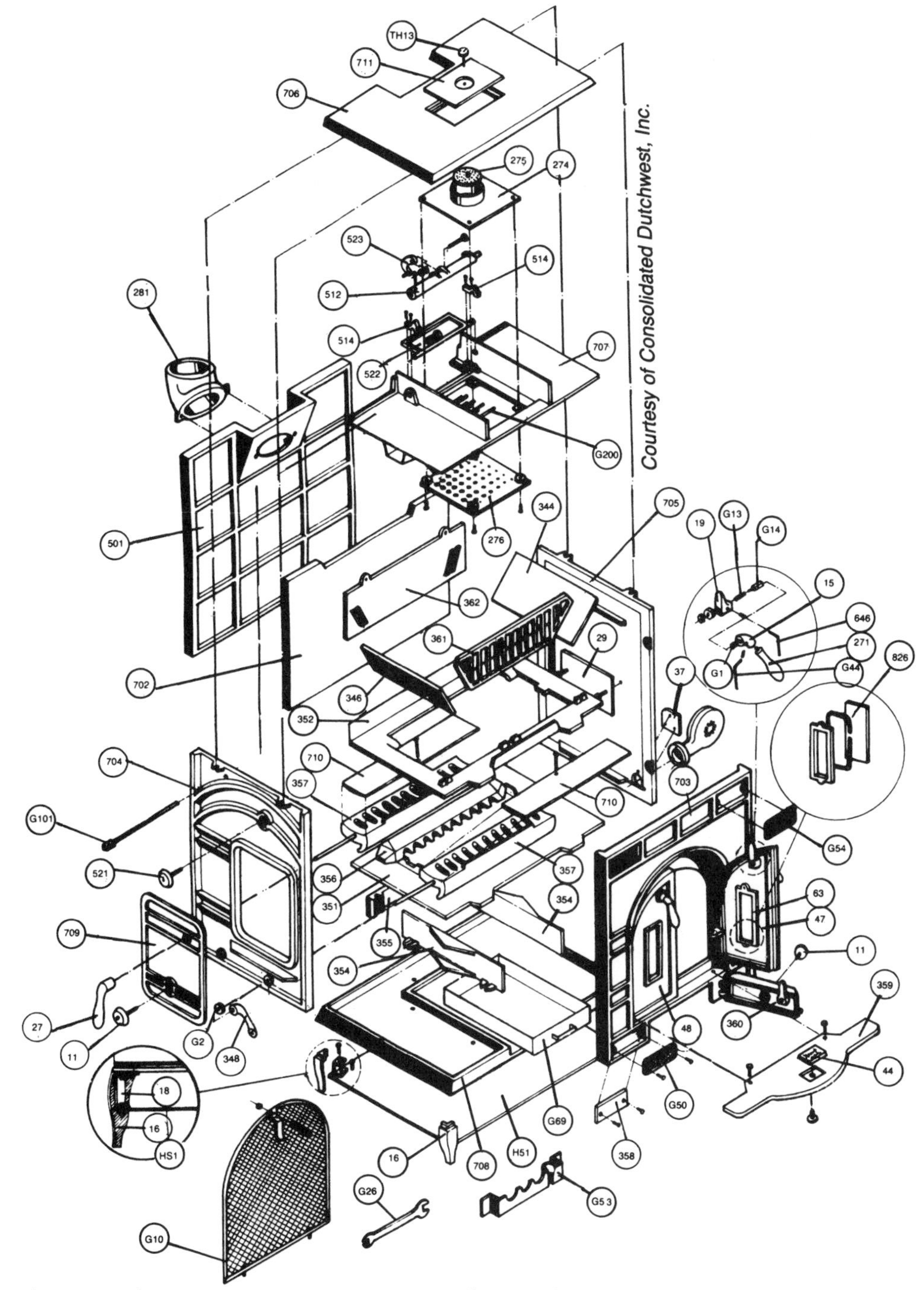

Fig. 10-12. The components of a freestanding wood stove.

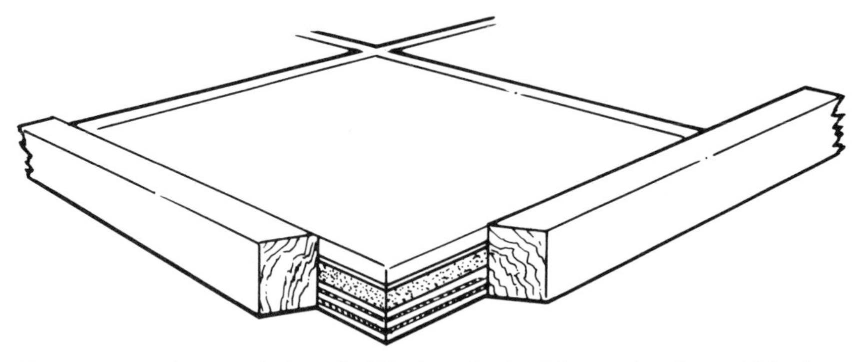

Fig. 10-13. A portable hearth. The hearth should extend at least 10 inches on each side of the stove and 18 inches on the loading door side. The hearth is made of a plywood base, half-inch mineral board, 22-gauge sheet metal, and a quarter-inch slate on the top. The edge is surrounded by a trim board. Adapted from a Consolidated Dutchwest drawing.

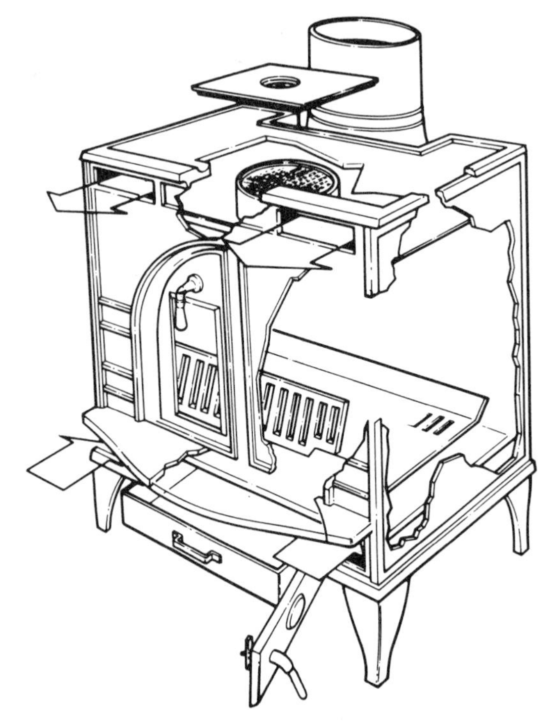

Fig. 10-14. A typical wood stove that shows all of the component parts. A catalytic combustor is visible at the top of the stove as a short cylinder with perforations in the top. Adapted from a Consolidated Dutchwest, Inc. drawing.

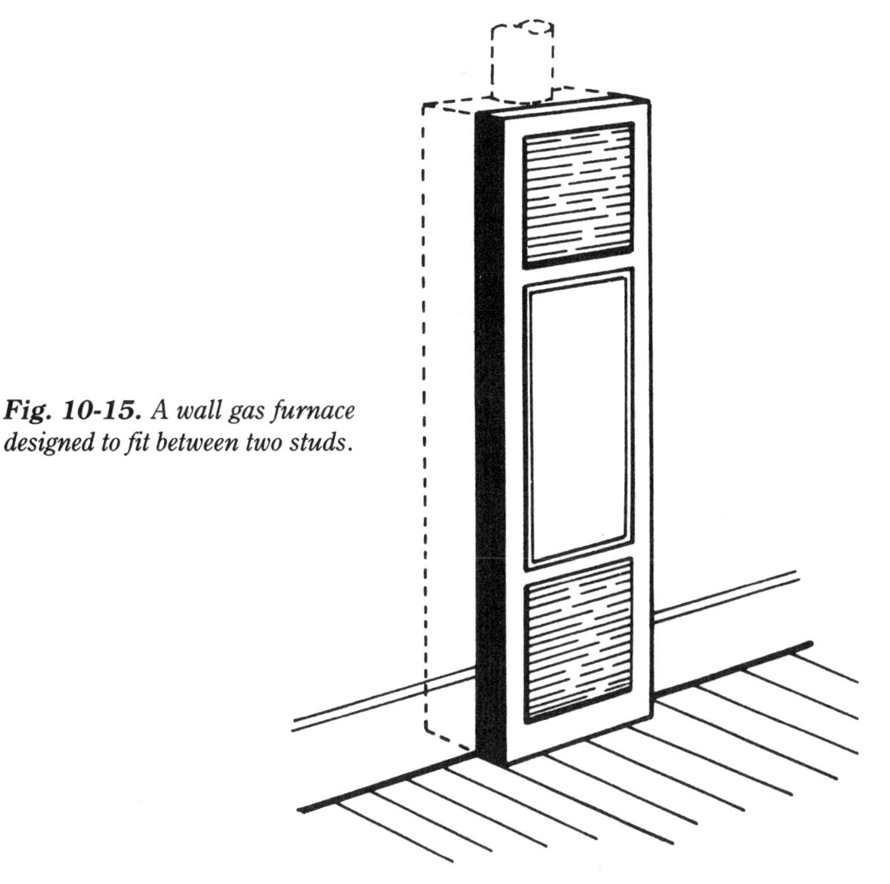

Fig. 10-15. *A wall gas furnace designed to fit between two studs.*

to load from the side. Experience can be your guide. Loading through either door works fine although longer logs can be loaded through a side door if your unit has one. Models FA207CL, FA209CL, and FA211CL are deeper than they are wide and thus load quite easily from the front. Starting a wood fire in a wood stove is not much different from a fireplace or a campfire. Follow these steps to build an effective fire:

Step 1. Make sure your stove is set up for wood burning. Your catalytic combustor should be set into the top of the stove in the round space provided just beneath the removable polished top. Remove the cast-iron strainer in the combustor space, it's only for coal burning and will interfere with the catalytic combustor. If your stove is a FA207CL, FA209CL, or FA211CL you also need to remove the drop-in coal front. If your stove is from the 224, 264, or 288 series, then you need to be sure all coal burning parts have been removed.They are two coal sides and the grate front. The grate front can be left in place for wood burning to keep logs away from the front doors (load from the side). The

FA267CL model also has a front grate that is useful in keeping wood off the front doors. All stoves come with grate covers that are found to marginally improve stove efficiency by approximately two to three percent. For more information on grate covers, see the assembly instructions that came with your stove. Use small pieces of wood—preferably soft woods—as kindling, use hardwood for your main fire.

Step 2. Open the bypass gate, and any flue damper(s). This is very important because it minimizes draft resistance and ensures your stove will not smoke when the doors are opened. If you attempt to open the doors while firing is underway and the dampers are closed, your stove is almost certain to smoke. This is dangerous. The enhanced draft experienced when the dampers are open also makes fire starting easier.

Step 3. Using paper, kindling, and larger pieces of wood, organize a normal wood fire. Be sure to use sufficient quantities of all materials so that the fire has an opportunity to catch.

Step 4. Before lighting the fire, be sure the side door (if you have one) is closed. Close the secondary air supply as well. Check again to make sure the bypass gate is open. If you have a flue pipe damper, be certain it is open as well. Open the primary air supply three turns. Light the fire and immediately close the main door(s). With only the primary air open, a bellows effect is immediately created which will quickly bring the fuel to a full burn. Let it burn three to seven minutes, then add additional larger wood. Within 10 to 15 minutes, the fire will be well established.

Sometimes when a chimney is cold, it will take it a minute or two to warm up sufficiently to draw. Unfortunately, this can lead your stove to "smoke" until it starts drawing. To alleviate this problem, try rolling up a couple of pieces of paper, and placing them on top of the fuel in your stove, then push them toward the stove back. Light these pieces first and close the front door(s).These will heat the chimney sufficiently to start it drawing. Through the glass window(s), you will be able to see when the chimney starts drawing the smoke away. Once it does, open the main door(s) and light the rest of the fuel from the bottom. Do not light the main bed of fuel until the chimney begins drawing, and if the rolled paper goes out without creating a strong draw, repeat the procedure.

Step 5. Once the fire is well engaged, close the bypass gate. This will activate the catalytic combustor, if one is used, and force the heat through the stove's baffling system. Once the thermostatic probe on the stove top passes 500 degrees, the combustor has begun to operate. To ensure its continuing operation, let the temperature approach 800 degrees before changing your air settings.

Step 6. Open the secondary air supply about one turn open. Close the primary air completely. The volume of fire will be immediately reduced, but the stove will continue to warm up. Continue to control the fire using only the secondary air. For a smaller fire, reduce the secondary air setting. For a larger fire, increase the damper setting. If this does not produce a sufficiently hot fire, then reopen the primary damper a small amount.

Step 7. If you have installed a stove pipe damper—you may want to partially or completely close it at this point. A partially or fully closed pipe damper reduces the volume of flue exhaust. With the damper at least partially closed, heat tends to be retained in the stove longer, while at the same time secondary combustion tends to be more complete. Note that even when the stove pipe damper is "fully closed," the pipe remains 30 percent open because of openings provided in the face of the damper and space around it. **Never open the stove when either the pipe damper or bypass gate is closed. Smoke will come out the doors if the dampers are closed when the doors are opened.**

Note about operating temperatures: Wood smoke activates the catalytic combustor, which causes the smoke, creosote, and other by-products of burning wood to ignite, resulting in a very high temperature in the combustor. Although operating temperatures between 1,400 and 1,600 degrees are common, the recommended operating temperature range is 1,200 to 1,400 degrees to prevent damage to the combustor. Temperatures of 1,800 degrees or higher will damage the combustor. High combustor temperatures are frequently experienced even during slow burning, cooler fires because this type of burning actually produces more by-products, which is fuel to the combustor. Also, the combustor probe set into the cook top will register very high temperatures even though the actual temperature of the stove body itself will be much lower. Typically, the stove body will be 300 to 600 degrees.

ADDING MORE WOOD

When you want to add more fuel, open the bypass gate (and the stove pipe damper if you have one) and wait 15 seconds, then open the door slowly. Add fuel. Split wood will more completely fill the firebox and reduce the frequency of reloading. If your wood is not seasoned (dry), it will burn more readily if split.

After closing the loading door, check to see if the thermometer is continuing to register at least 500 degrees, the minimum temperature required for catalytic activity. Frequently, the temperature will drop below 500 degrees, particularly if the loading door is open a long time. You might want to increase the secondary air setting to prompt the fire. If the fire is very weak, you can open the primary air and/or the bypass to increase the draft. This should be done only for very brief periods, however, as the stove might over-fire, sending significant amounts of lost heat up the chimney while wasting fuel.

Removing wood ash: A natural residue of wood burning is wood ash. Ash can be left to accumulate in the stove without risk to the stove, but it should be removed if its accumulation is causing a mess. Accumulations of wood ash seem to enhance catalytic activity, because wood ash is mainly charcoal, and charcoal provides a high percentage of the gases that the combustor burns most readily. Typically, during wood burning, the ash drawer needs to be emptied once every one to three days. **Wood ash should always be disposed of in a safe manner.** Use a noncombustible, tight-sealing container. If you keep ash in the house before disposing of it, be sure it is placed on a noncombustible surface with no combustible material in contact with it. Wood ash may be used as a fertilizer, so if you're a conscientious gardener you might want to save it.

BURN TIMES AND LONG BURNS

When firing your stove, the volume of heat and the length of your burn time will vary according to the size of your stove, the damper setting, the volume of fuel loaded, and the quality of your fuel. In general, the normal operating temperature of a stove is about 300 to 750 degrees (temperatures on the front and side of your stove). The probe thermometer on the stove top, however, during catalytic combustion, will register about twice the temperature of the stove body. During coal burning, the temperature recorded on the top approximates the temperature of the entire stove body.

CATALYTIC COMBUSTION

The catalytic combustor is essential for high efficiency and clean burning. When wood is burned, only a fraction of its energy is turned into heat. The rest goes up the chimney in the form of smoke. It takes a very high temperature of 1100°F to burn the smoke and turn it into productive heat. In noncatalytic airtight stoves, some of this smoke burns during high burns and where additional combustion air is provided. This process is ineffective during medium and low burns, however, because firebox temperatures are not high enough to reach the necessary combustion temperature. These stoves also produce a large amount of creosote during long burns.

The catalytic combustor is coated with precious metals that cause the smoke to burn at only 500-500°F, meaning a more complete combustion process that gives you more heat and a cleaner burn. Even during long burns, when the firebox temperature decreases, the combustor feeds on the smoke to maintain the 500-600 degrees combustion temperature. If properly used, the result is longer burns, more heat per pound of wood, and less creosote.

COMBUSTOR USE

In order to activate the combustor, the bypass gate must be closed to force the smoke up through the combustion chamber. Because this reduces the draft, you should be careful not to close the bypass gate too soon. When starting a fire, even if the temperature probe reaches 600 °F within a few minutes, you should let a strong fire run about 15 to 20 minutes before closing the bypass gate and activating the combustor. If you close the bypass gate too early, you might put out the fire or deactivate the combustor. Thus, a cold stove should be warmed up well before closing the bypass gate. Keep in mind that when opening the front or side loading doors, make sure to first open the bypass gate and optional stove pipe damper to prevent heavy smoking.

The combustor air supply feeds air to the area just below the combustor between the combustor and baffle for all models except model FA224CCM. The combustor's efficiency is enhanced by fully opening the combustor air dial damper high up on the side of the stove during medium and high burns. The preheated air helps burn off the heavy volume of smoke more completely. The dial damper should be closed during low burns (8 hours or more), because additional air is not needed to create secondary combustion of smoke and by-products in the combustor.

A ceramic combustor enhances your stove's performance, but there are certain characteristics that you should be aware of and certain precautions you need to take:

1) Burning fuels other than natural wood will shorten the combustor's life substantially. If you burn coal or wood that has been treated in any way, the combustor should be removed from the stove. Otherwise, ash and other chemicals will plug up the combustor or cause the precious metal coating to deteriorate. Colored or coated papers, or papers printed with colored inks, will have the same effect if burned in large quantities. If you forget to remove the combustor when burning coal, and ash has built up, try burning a few high-wood fires with the combustor in position to get the ash coating off.
2) The stove should be operated at probe temperatures between 1200 to 1400 degrees. Temperatures over 1800 degrees will damage the combustor.

3) If you notice creosote building up in your chimney or pipe, you might be operating your stove in a way that produces creosote or your combustor could be plugged up. Certain woods leave an ash film on the combustor, plugging it up. Periodic inspection of the combustor will solve this problem.
4) The combustor might glow during the first quarter to one-third of the burn cycle, but it does not have to be glowing for it to be working.

CREOSOTE

When wood is burned slowly, it produces tar and other organic vapors that combine with expelled moisture to form creosote. The creosote vapors condense in the relatively cool chimney of a slow-burning fire. As a result, creosote accumulates on the flue lining. When ignited, this creosote can cause an extremely hot and dangerous fire. With the catalytic combustor and combustor air to burn off wood smoke and by-products, creosote buildup will be reduced by as much as 90 percent. Many people have gone through a number of seasons with little creosote buildup. Nonetheless, greatly reducing creosote buildup varies with different installations and firing habits. Installations with poor drafts are more likely to create creosote because wood smoke moves very slowly up the chimney and is more likely to cool and condense there. Stoves set for long burn times (especially those with a poor draft) can also contribute to creosote during the last few stages of the burn. This small amount of smoke might not be enough to keep the combustor activated. You should monitor your stove at the beginning to see how long your stove can keep the combustor activated and how much fuel it takes. Typically, in bigger models, the combustor remains active up to about 12 hours and about five to eight hours in smaller models. You should keep an eye on the probe thermomenter on top of the stove. If the temperature falls below about 500 °F, and you keep the stove closed up, creosote will build up. In this case, rebuild the fire or open up the bypass gate to create a stronger draft.

Even when using a catalytic combustor, your chimney connector and chimney should be inspected regularly, at least once every two months in the heating season—more often if experience dictates—to determine if creosote buildup has occurred. If creosote has accumulated, it should be removed to reduce the risk of a chimney fire. Purchase a few chimney brushes to make cleaning easier.

OPERATING RULES

Take the time to read all of the information provided with your new wood stove. It will make for a safer installation, safer operation, and more efficient performance. The following is a brief list of important things to remember.

1) Anyone operating the stove should read the operating instructions for a clear understanding of its operation.
2) Your installation should be checked by a local building inspector or, in the event none is available, by a reputable chimney sweep. Many communities require inspections, so always contact your local officials for applicable requirements before using your stove.
3) If installing in a fireplace, the fireplace chimney should be lined and a direct stovepipe connection made between the stove and the chimney. Inspect your chimney first for blockage or signs of deterioration and clean if necessary. Do not remove bricks or mortar from the fireplace. An older, unlined chimney, where the masonry was installed against or very close to wood studs or other combustibles can be a very serious fire hazard. If you have any doubts about your chimney, we strongly urge you have it inspected by an expert (building inspector, etc.).
4) **Do not connect a unit to a chimney flue serving another appliance.**
5) Always keep the area around the stove clear of furniture, drapery, and other objects. A 48″ or greater clearance is recommended. Be sure drapery and loose papers cannot blow closer than this.
6) When starting a fire, no liquid fuel, such as kerosene, gasoline, engine oil, or lighter fluid should be used. Approved fuels are wood, bituminous (soft) coal, and anthracite (hard) coal. No other fuels should be used. Cannel coal is not acceptable (its gases are potentially explosive). Neither are parafin logs. Do not burn garbage in your stove.
7) Your first few fires should be small to medium in size and wood should be your fuel. Cast iron needs a few milder fires to "cure" (reach its maximum strength), so you should not build an intense fire the first few times. This could crack a casting. Build up each succeeding fire a little hotter. After four or five fires, your stove will be "cured." In the

case of smoke coming from surfaces of the stove during your first fires, see the Troubleshooting instruction #10. Use your temperature gauge and air inlet controls as guides when curing the stove.

8) When starting a fire or reloading, the bypass gate must be open. To open, slide a removable door handle onto the bypass gate handle shaft and rotate it. If you have a pipe damper or an adapter/damper, they must also be open. If the door(s) are opened while either damper is closed, the stove will smoke heavily. **Do not under any circumstances open the door(s) while these dampers are closed.** Open the doors a crack for a few seconds before opening all the way. This will help avoid back drafts. Wait a little longer before opening the doors with a coal fire.
9) On stoves having both front and side doors, you should never open both doors at the same time. Otherwise, cross currents could lead to combustion products spilling from one of the doors.
10) Do not store fuel close to your stove. Four feet or farther is the recommended distance. Do not install a stove in a garage that is usually closed. The presence of a stove near combustibles such as gasoline, engine oil, etc. is a very dangerous situation.
11) **Do not operate the stove with the ashdoor open, this will lead immediately to extreme over firing.**
12) Do not over fire your stove. A glowing stove or chimney is a sign of over firing. Immediately close all dampers and monitor the fire until it has returned to a normal level. Do not attempt to overload your stove as this could lead to an over firing situation. Never run your stove so hot that it glows.
13) If your stove is equipped with a fire screen for open hearth wood burning always clip this in place if you want to operate the stove with the front door(s) open. **Do not operate your stove with the front door(s) open unless the fire screen is firmly in place. Do not burn coal with the front door(s) open. Be sure to open the bypass gate and any flue damper before putting the screen in place.**
14) Remove your catalytic combustor when burning coal or treated or painted wood. In its place set the cast-iron strainer. When converting from coal to wood operation remember to remove the strainer and set the catalytic combustor in place. It increases efficiency by more than a third. Remove/replace the combustor only when the stove is cold. With the combustor in place, do not burn colored or coated paper (such as catalog stock or paper printed with colored ink).

Never burn pressure-treated wood in any stove (with or without a combustor), as it contains a toxic chemical, copper chromium arsenate, a form of arsenic. Burning plywood and chipboard or particle board will also produce toxins.

15) Do not operate the shaker grate(s) with the ashdoor open. Hot coals might fall out. To prevent the spread of dust, try closing the loading door(s) as well.
16) When using the rotating coal grates for coal burning, remember the rotating sections have a range of 90 degrees. For ordinary shaking, rotate them 45 degrees. Rotate them 90 degrees only when you wish to dump the coal bed. Rotating more than 45 degrees will probably cause the grate to jam during firing. If the rotating coal grate(s) jam during ongoing operation, wait 45 minutes to an hour. In most cases, the trapped coal will be consumed in the fire and you will be able to free up the grate.
17) When disposing of ash, exercise extreme care. Ash frequently contains hot coals. Dispose of them in a noncombustible container with a tight-fitting lid. Do not dispose of them near a house. Remember, coals can remain hot for a long period of time, so they should never be dumped close to combustible materials. Do not store them in a garage—the possible presence of flammable gases is a very dangerous situation.
18) Inspect your installation periodically for any evidence of deterioration.
19) Most stoves come with removable (slide off) door handles. They should always be removed when the stove is fired, because they will become very hot and could cause burns. Small springs are provided to fit inside the handles so they will fall off in case you forget to remove them. A wall-hanging bracket, which can be secured to a wall with two screws, is provided to hold them when the stove is in use. The removable handles can also be used to adjust the dial dampers, open or close the catalytic bypass gate, and remove the fire screen. It is a very good idea to have a pair of insulated gloves for use when loading a stove so you won't accidentally burn your hand.
20) When wood is burned slowly, it produces tar and other organic vapors, which combine with expelled moisture to form creosote. The creosote vapors condense in the relatively cool chimney of a slow-burning fire. As a result, creosote accumulates on the flue lining. When ignited, this creosote can cause an extremely hot and dangerous fire. The chimney connector and chimney should be inspected regularly, once every two months in the heating season—more often

if experience dictates—to determine if creosote buildup has occurred. Using a catalytic combustor greatly reduces creosote build-up (as much as 90 percent), but you should still regularly inspect your chimney to be certain it is clean. If your stove is not already equipped with a catalytic combustor, you might want to purchase one. If creosote has accumulated, it should be removed to reduce the risk of a chimney fire. Consider purchasing a chimney brush to make cleaning easier.

21) When coal is burned, it produces fly ash, which can accumulate in the chimney, particularly around elbows. You should inspect your chimney every two months (more often if experience dictates) for fly ash and remove it if necessary.

22) When coal is burned, carbon monoxide is produced. Normally, this is drawn up the chimney like the smoke from a wood fire. Unusual circumstances, such as an obstruction or leak in the chimney, however, might cause this gas to leak into the house. **It is extremely dangerous.** While carbon monoxide is odorless, coal fires give off other gases known as aldehydes. Aldehydes have a distinctive "sour" odor. Thus a sour odor coming from your stove or chimney during a coal fire means carbon monoxide might be present. The first physical symptoms of carbon monoxide poisoning will normally be a severe headache, dizziness, and possibly an upset stomach. If sour odors or these symptoms are noticed you should (1) immediately open the doors and windows of your house to let in fresh air; (2) shut down the stove by closing the dial dampers all the way; (3) investigate further to determine whether the stove or chimney is leaking; (4) check to see if the chimney has become blocked; (5) call in a qualified installer for an inspection if the source of the problem cannot be determined; and (6) correct the problem before burning again.

23) Should you ever experience a chimney fire, you should immediately warn everyone in the house. Children, elderly, and infirm persons should leave immediately. If the fire is serious or has spread beyond the chimney, everyone should vacate the house immediately and the fire department be called. Only if it is safe to remain, close all the stove's doors and dampers to deprive the fire of oxygen. Do not throw water on the stove as this could cause it to break. It is recommended that you purchase fire extinguishers for your home if you do not already have them. A flare type chimney fire extinguisher can be lit and tossed in the stove. This will create a smoke which consumes oxygen, smothering the fire. If you have a barometric damper, block

the damper door closed so it does not let in more air. If it is safe to remain, move all of the furniture away from the stove and chimney (including upstairs if the chimney passes through a second floor). If this does not stop the fire, leave the house and call the fire department. It is also a good idea to hose down the roof to prevent sparks from spreading the fire.

11

Servicing Heating and Air Conditioning Systems

THERE ARE A GREAT MANY HEATING EQUIPMENT REPAIRS THAT A HOMEOWNER can easily and safely accomplish. By following the easy, step-by-step instructions given in this chapter, you will seldom need to call a professional for help. Be sure you have read Chapter 1 on Safety Precautions first, before you begin.

GAS-FIRED FURNACES OR BOILERS

When the components of a gas-fired, warm-air furnace or hot-water or steam boiler are described individually and explained, repairs and adjustments are relatively easy. The standard, warm-air furnace consists of the following: the heat exchanger; the air circulating fan; the burner; the casing; and the controls. See Figs. 11-1 through 11-8.

The Heat Exchanger

The heat exchanger is a semi-sealed firebox housing the flame part of the burner. Most modern heat exchangers have a box section on the bottom, open on the front. The burner sits in this section. Each section of the burner consists

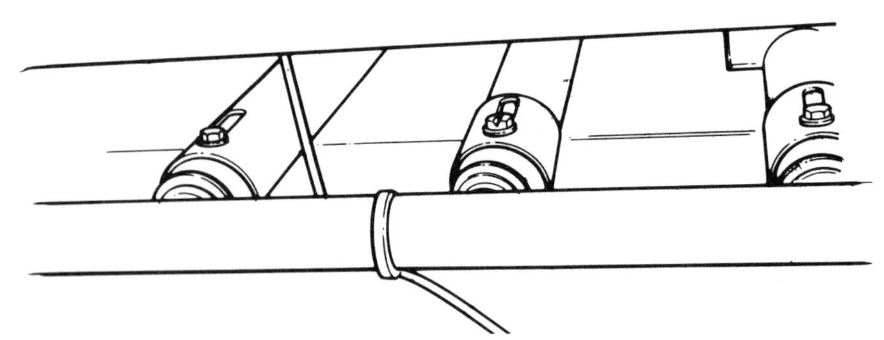

Fig. 11-1. The mixing tubes of a gas furnace. The manifold extends across the front. Gas orifices are screwed into the manifold, and the mixing tube ends fit over the orifices. The setscrews in the slots on each tube are for adjusting the shutter to provide optimal ratio of gas and air for proper flame appearance.

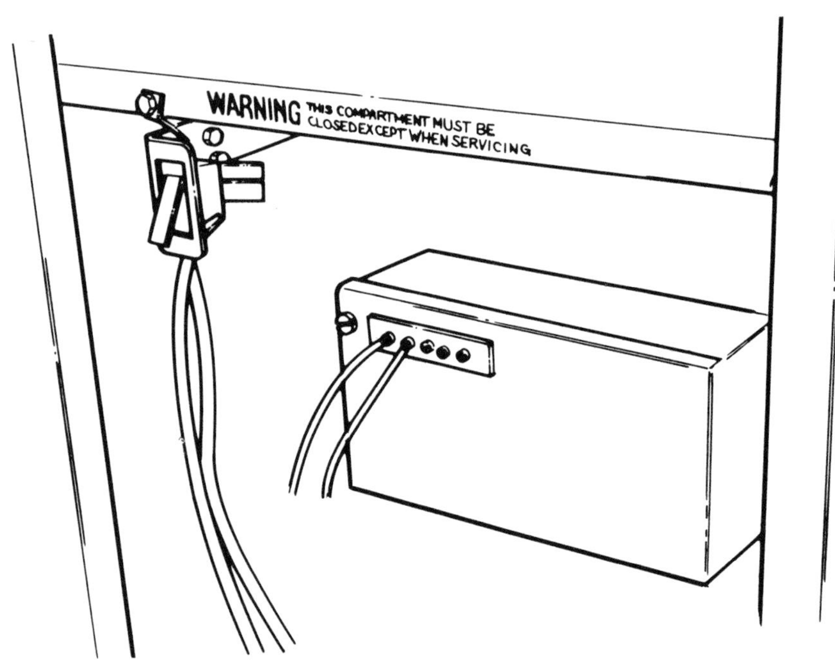

Fig. 11-2. A control box and interlock switch that does not allow the fan to run unless the compartment door is in place.

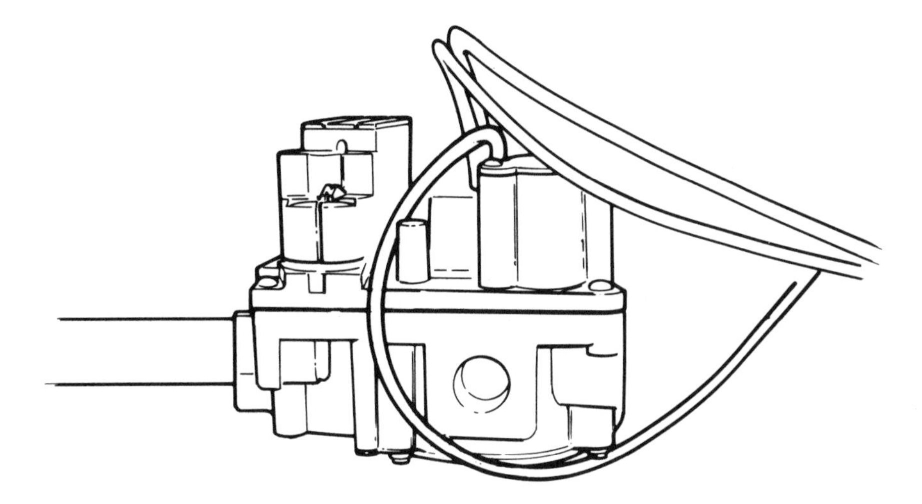

Fig. 11-3. *An automatic gas valve on a furnace. The line looping down is the safety control tube to the thermocouple at the pilot.*

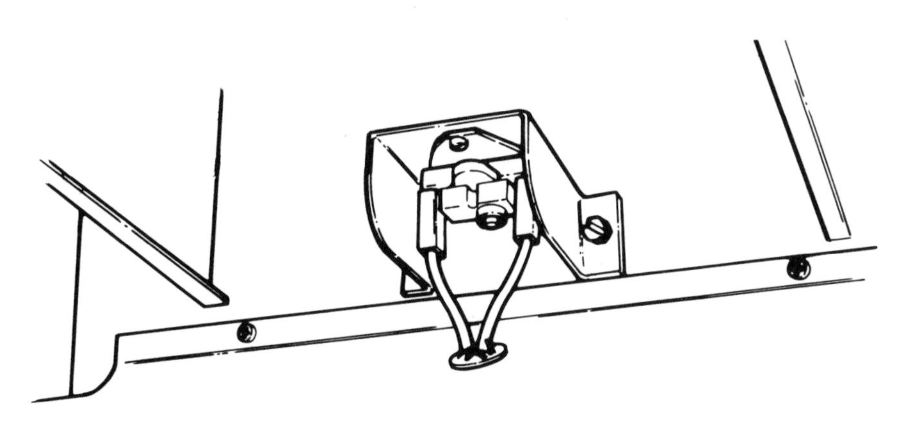

Fig. 11-4. *This safety switch does not allow the burner to operate if the flue or chimney is blocked or plugged.*

of a tube with a row of small holes where the gas flame comes out. These tubes are connected together at the front by a manifold that supplies gas and air to the gas flame. Each tube, with its flame, sits directly under a sealed passageway, usually four or more, which terminate at another sealed box at their tops. This is connected to the flue pipe (smoke pipe) at the top rear of the furnace. See Figs. 11-8 and 11-9.

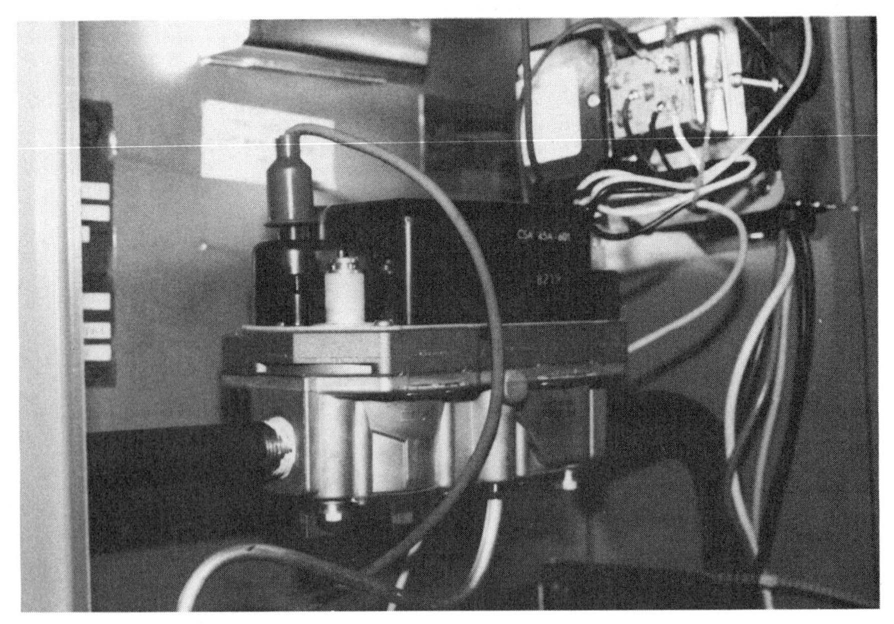

Fig. 11-5. *An automatic gas valve. A transformer and terminal board are shown in the upper right.*

Fig. 11-6. *A bank of three burner heads. The pilot and thermocouple are shown to the right of the center head. The line curving beneath is the gas line to the pilot burner.*

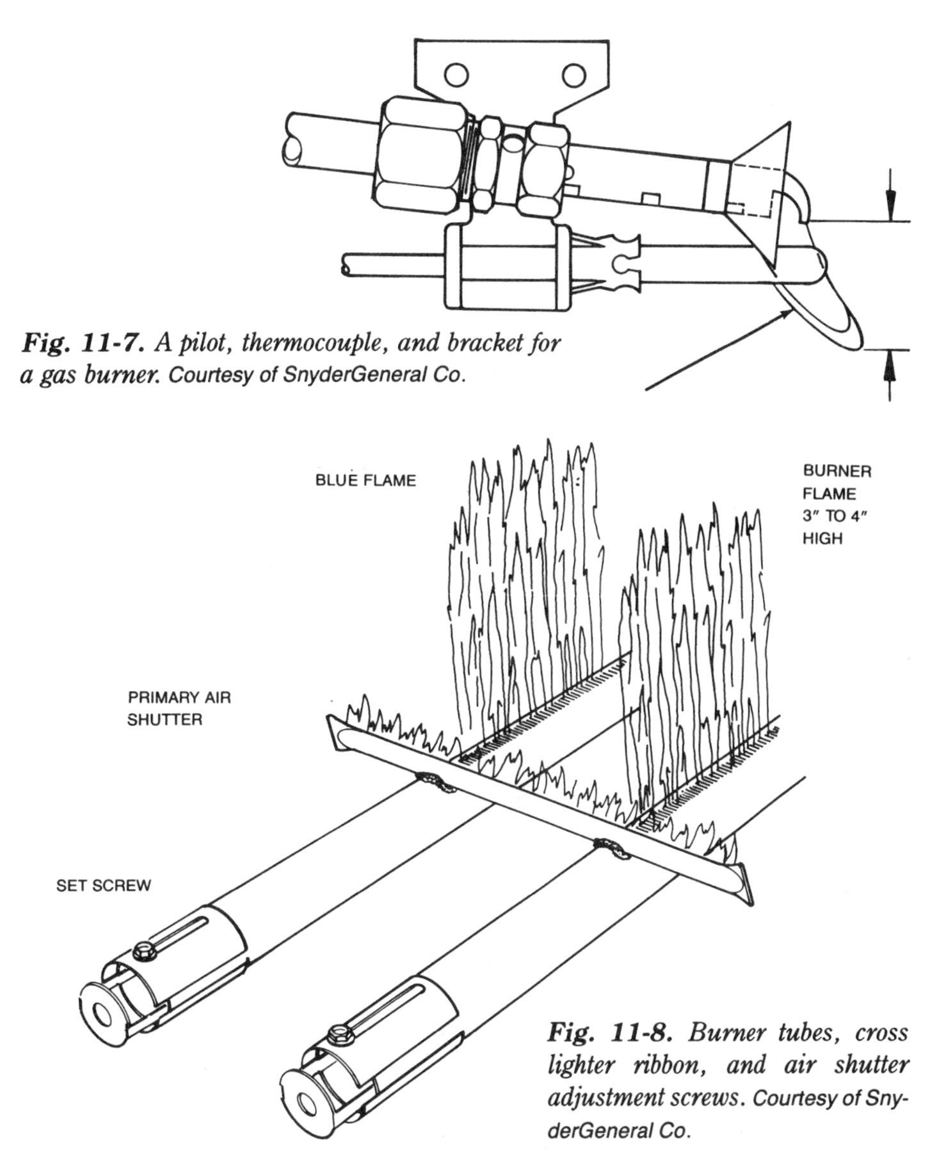

Fig. 11-7. A pilot, thermocouple, and bracket for a gas burner. Courtesy of SnyderGeneral Co.

Fig. 11-8. Burner tubes, cross lighter ribbon, and air shutter adjustment screws. Courtesy of SnyderGeneral Co.

The vertical passageways conduct the hot gases from the burner up and out of the flue pipe. As the burner operates, the air-circulating fan forces air around the passageways and the top and bottom boxes, which are all sealed together, and picks up the hot air and delivers it throughout the house. Once the air is returned to the furnace fan, it is pulled through air filters to remove dust and dirt. All of the component parts are enclosed in an outer casing with doors so that the burner and the fan can be accessed. See Figs. 11-11, 11-9, 11-12, and 11-13.

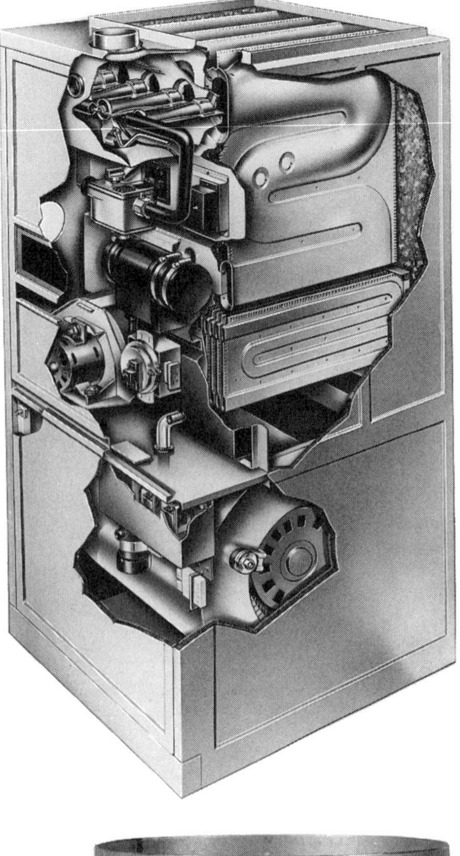

Fig. 11-9. A cut-away view of the furnace shown in Fig. 11-7. Courtesy of Carrier.

Fig. 11-10. A flue damper to close off the flue pipe when the burner is off. Courtesy of Energy Vent, Inc.

Fig. 11-11. A typical forced, warm-air gas furnace. Courtesy of Carrier.

Fig. 11-12. A gas pilot burner with the ignitor on the same bracket. Courtesy of Honeywell Inc.

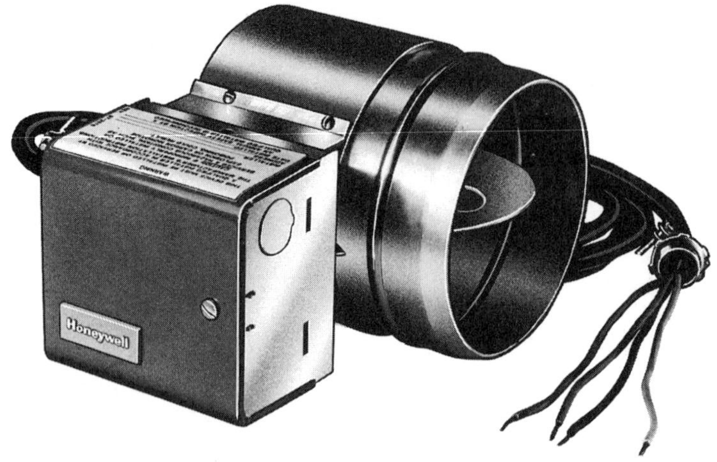

Fig. 11-13. *A vent damper for the flue pipe can save heat when the furnace is off. This one includes wiring for ease of installation.* *Courtesy of Honeywell Inc.*

The Air-Circulating Fan

The air-circulating fan is generally in the bottom section under the burner compartment. Fans are furnished with two types of motor mountings. The external type motor is mounted either on top of, or on the side of, the fan housing, and drives the fan by means of a fan belt and pulleys. NOTE: This belt must not slip nor must not be too tight. Make sure the furnace electric switch is turned off then push down on the belt midway between the pulleys. One inch movement is about right.

This type of motor is mounted on a hinged base. Opposite the hinge is a bolt threaded through the base with a locking nut snugged against the base to hold the bolt in position. The end of the bolt has a rubber pad that pushes on the fan scroll to tighten the belt. This arrangement gives a permanent, positive adjustment for the belt tension. The external motor has oil cups or holes for adding lubrication. The fan shaft also has lubrication points, cups, or holes. The fan oil cups should be filled twice a year. If there is no reservoir, holes will not take very much oil and might need oiling three times a year. Oil motors twice a year; use *only* four to six drops each time, in each opening. Too much oil will run down inside the motor and get on the internal starting switch points. This oil, along with dust and dirt, will foul up these points and prevent the motor from starting. **Caution**: Oil used for motors and bearings on fans **must be non-detergent oil.** Do not use automotive engine oil! In rare cases, you might be able to locate engine oil that is nondetergent. The words nondetergent must be printed on the container. Many hardware stores will carry oil specially packaged for these bearings. This is not the ''Three-in One'' type oil and will say ''heavy duty oil for electric motors.'' Automotive type oil will extensively damage these types of bearings.

The second motor type is mounted inside the fan squirrel cage. Its mounting bracket has three legs that are bolted to the fan housing. The squirrel cage mounts directly on the motor shaft and has no shaft and bearings of its own. Many of these motors have sealed bearings and do not need lubrication. If there are oil holes or cups, oil as any external motor, see Fig. 11-9.

The Burner

The gas burner seldom has any moving parts and operates similar to the burner on a gas range. Leaving the gas valve is the manifold, a pipe about one inch in diameter extending across in front of the heat exchanger. The burner tubes plug onto the manifold over orifices—small pipe plugs with a drilled hole, sized to permit a certain amount of gas to flow out and into the burner part. The burner end flares out and has an air adjustment plate over the large end to adjust the amount of air needed for complete combustion.

Controls

The furnace controls for gas fuel are quite simple. The main control is the gas valve combined with the pilot control valve. Normal start-up requires that the main valve and pilot valve be closed and the furnace shut off for five minutes. Then the control knob on the gas valve combination is turned to the ''pilot'' position (you might have to push down on the knob on some models before turning). Push the knob down and hold a match to the pilot burner. You will probably need more than one match to get the pilot to stay lighted, especially if any of the gas piping has been opened and reconnected. Air gets trapped in the gas piping and must be pushed out through the pilot line before gas is available to the pilot. This might take a few minutes as the pilot orifice is very small. Have patience and more matches on hand. See Figs. 11-14 and 11-15.

When the pilot flame is burning steadily, the knob can be turned to ''main burner.'' The burner will light if the thermostat is set higher. Figure 11-16 shows an intermittent pilot control.

Another control is the limit control, which will shut the burner off if for any reason the furnace overheats. A broken fan belt, a fan motor that does not start, a clogged filter, or hot-air registers shut or restricted can cause the furnace to overheat.

The control operating the air-circulating fan is sometimes in the same case as the limit control or as a separate unit. This control turns the fan on when the furnace is heated up to about 130° and shuts the fan off at about 90°.

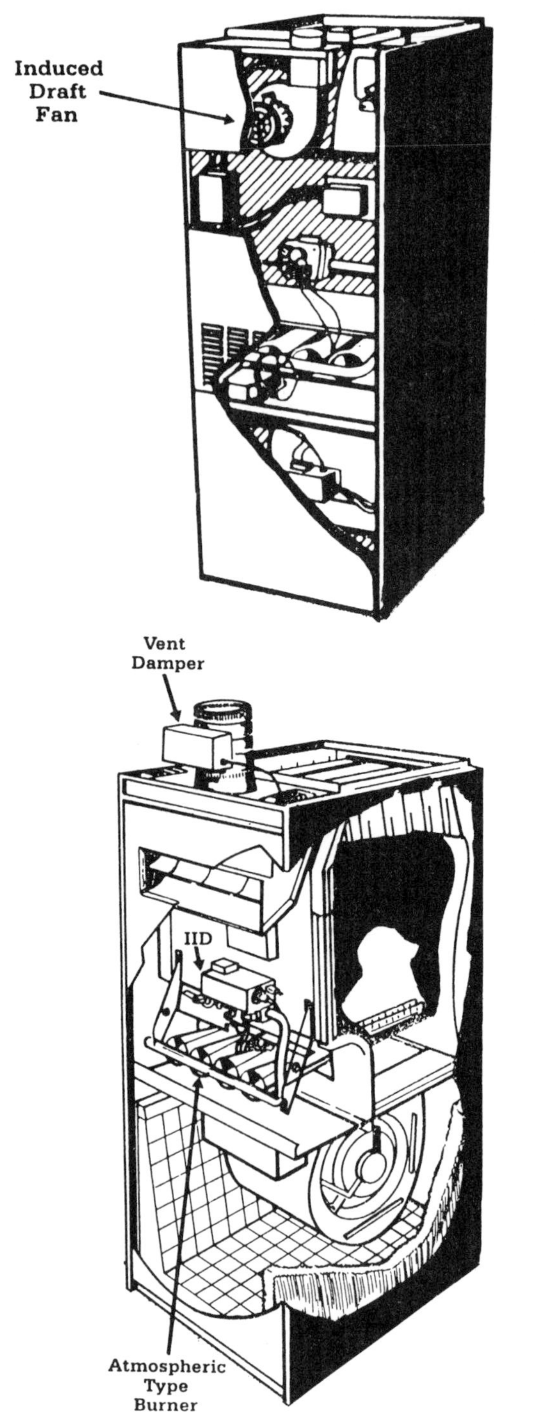

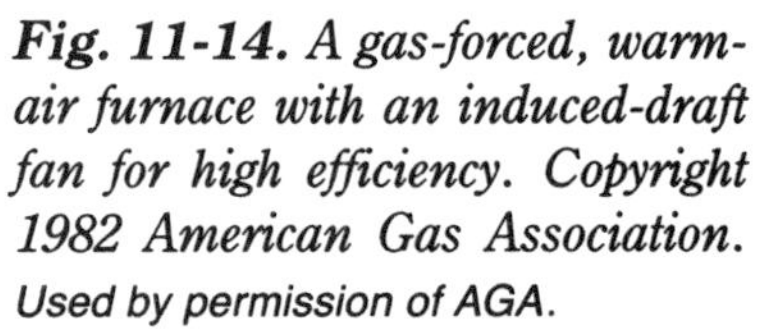

Fig. 11-14. *A gas-forced, warm-air furnace with an induced-draft fan for high efficiency. Copyright 1982 American Gas Association.* *Used by permission of AGA.*

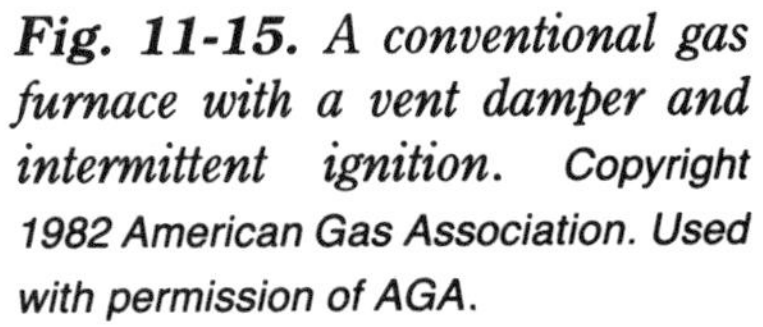

Fig. 11-15. *A conventional gas furnace with a vent damper and intermittent ignition.* *Copyright 1982 American Gas Association. Used with permission of AGA.*

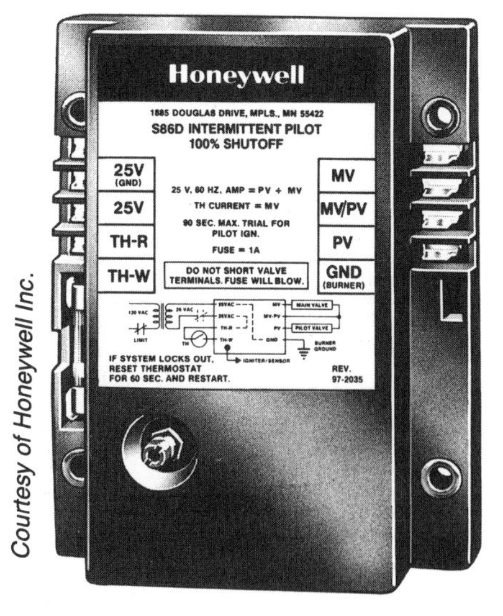

Courtesy of Honeywell Inc.

Fig. 11-16. *An energy-saving intermittent pilot control that lights the pilot when it receives a call for heat.*

This prevents the fan from blowing cold air. Air filters are necessary to filter dirt from the air being circulated. Filters should be replaced or cleaned, if cleanable, at least two times during the heating season. Change them on fall start-up and in late January.

Hot-water boilers are fired similar to forced air furnaces, only the controls are different. There is a boiler feed-water valve, which maintains the proper amount of water in the system, a relief valve to relieve excessive water pressure in the system, an operating water temperature control that directly controls the burner operation (this is different from the hot-air furnace as boiler water is kept at constant temperature of about 200°). The temperature in the house is maintained by the room thermostat. The thermostat cycles the water circulating pump, keeping hot water in the radiators or baseboard radiation. On some systems, a control can be used to keep the circulating pump from operating until the boiler water has reached about 150°, similar to the furnace fan control. A low water cutoff prevents the system from operating with a low water level.

In remote areas and other locations, the only gas available is LP Gas (Liquified Petroleum Gas). This gas, which is supplied in tanks, is similar to natural gas except that smaller orifices are used and the pressure regulator is different to accommodate the higher pressure of the liquified gas. **A very important caution**: Because LP Gas is heavier than air or natural gas, any leaking gas will actually "flow downhill" as a liquid would and accumulate in a low place. If ignited, LP Gas will explode violently. Once again, **ANY LEAKAGE WILL PRODUCE AN EXTREMELY VOLATILE, EXTREMELY DANGEROUS SITUATION!** I recommend that any repairs be done by a trained serviceman. Minor maintenance, such as changing filters and oiling motor and fan bearings can be done by the homeowner.

GAS FIRED, HOT-WATER HEATERS

Gas-fired, hot-water heaters are the most common type of heating system. The controls are similar to those on a gas furnace. The main control has the main gas valve, the pilot gas valve, and the thermostat (see Fig. 11-7). Also built into the case is a high limit control (energy cutoff) that shuts the gas off in case of excessive water temperature. In addition, in the top of the tank or near the top of the side, there is a temperature and pressure (T & P) relief valve. In case of high temperature or pressure, the valve opens and discharges water from the tank, relieving both conditions. **Note**: This is an important warning. **Any** mechanical device can fail to perform its function at any time without any warning. All persons should be aware of this and act accordingly!

Fig. 11-17. The Honeywell "Round" thermostat is a very reliable model that includes a mercury tube to make and break contact.

OIL-FIRED FURNACES OR BOILERS

Controls used with an oil burner are different from gas burner controls. Until a few years ago, the "stack relay" was used. This control, mounted on the flue pipe with a sensor inserted into the flue pipe, monitors the burner operation by sensing a rise in stack temperature, which can detect that a flame is established. If no temperature rise is sensed, the control locks out the burner. This is the burner safety control.

The operating cycle begins with the thermostat. The thermostat (on a hot-air furnace) or a boiler water control (on a boiler) starts the oil burner and establishes ignition, or an electric spark. (See Fig. 11-17) When the burner starts, oil is sprayed into the combustion chamber. The spark formed between the ignition points is also blown out into a half circle. This extension of the spark contacts with the oil spray and the oil is ignited. All this happens immediately and the flame is established. If no flame is established within ninety seconds, the relay locks out. This can happen when there is no ignition, no oil spray, a weak spark, a plugged nozzle, a defective oil pump, or the oil and air ratio are wrong (too much air). Excessive air causes a white flame rather than an orange-colored flame. This white flame has an acrid odor. This can be corrected by closing the air shutter slightly to produce an orange flame with *slightly* smoky tips. Figure 11-18 shows a cut-away view of an oil burner nozzle, and Figs. 11-19 and 11-20 show the component parts of an oil burner.

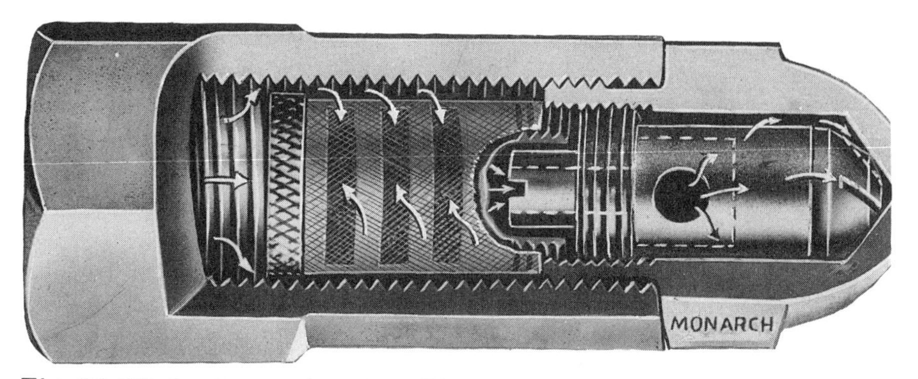

Fig. 11-18. *A cut-away view of an oil burner nozzle mounted in a nozzle holder. The path of the oil through the nozzle can also be seen.* Courtesy of Monarch Manu facturing Co.

Because the ignition points are behind and out of the oil spray, they stay clean. Some designs keep the ignition on during burner operation (constant ignition); others shut off ignition when the flame is established (intermittent ignition). This choice depends on the make of burner or specific conditions of the furnace and burner in question. Poor flame retention is one reason to use constant ignition. The constant spark will re-ignite the burner if the flame goes out during operation. See Fig. 11-22.

Newer burners have flame retention features that make them even more efficient and use an electric eye instead of the stack relay. The electric eye is mounted on the burner in a case or in the blast tube itself. The eye "sees" the flame and allows the burner to continue to operate. If a flame is not established, this control locks out the burner the same as the stack relay control does. Both of these controls are safety controls.

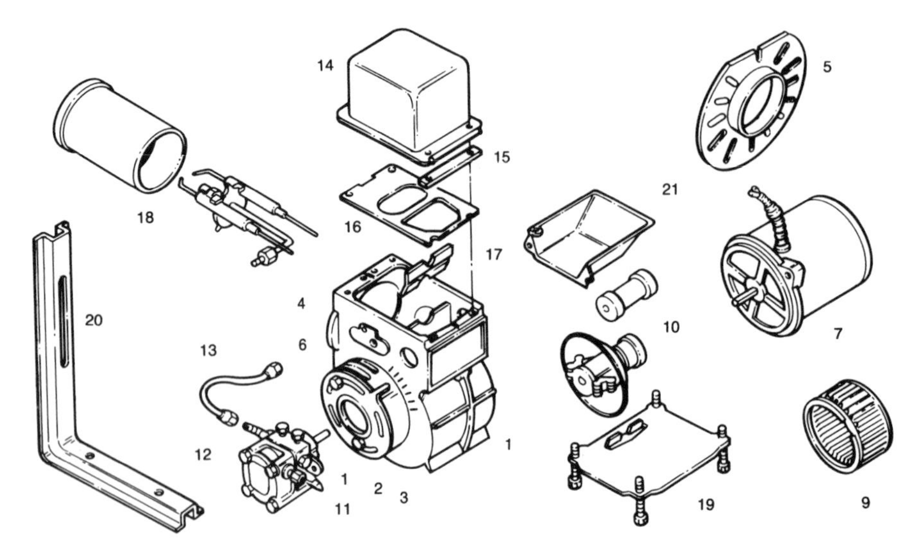

Fig. 11-19. The components of an oil burner, including part names and part numbers. Courtesy of Beckett Co.

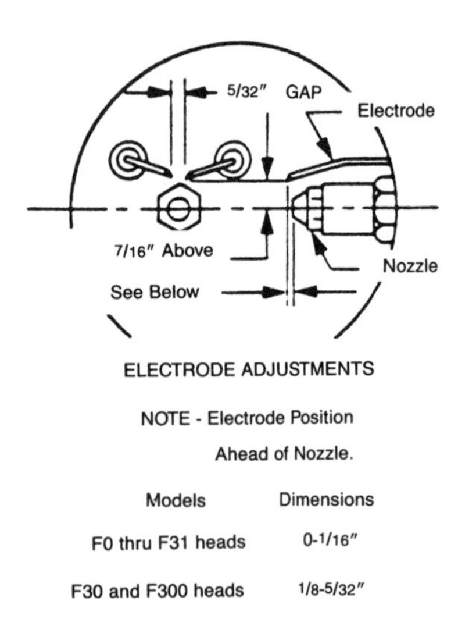

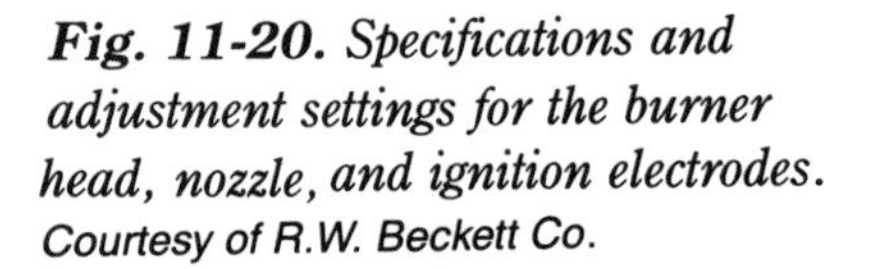

Fig. 11-20. Specifications and adjustment settings for the burner head, nozzle, and ignition electrodes. Courtesy of R.W. Beckett Co.

Electric Furnaces or Boilers

Electrically fired furnaces and boilers do not need a chimney and are very clean, but are *very* expensive to operate. Installation costs are cheaper than either gas or oil, but operating costs are usually double other fuels. Additional safety controls prevent energizing the heating elements when either the fan is not running on the hot-air furnace, or the boiler has low water or has no water.

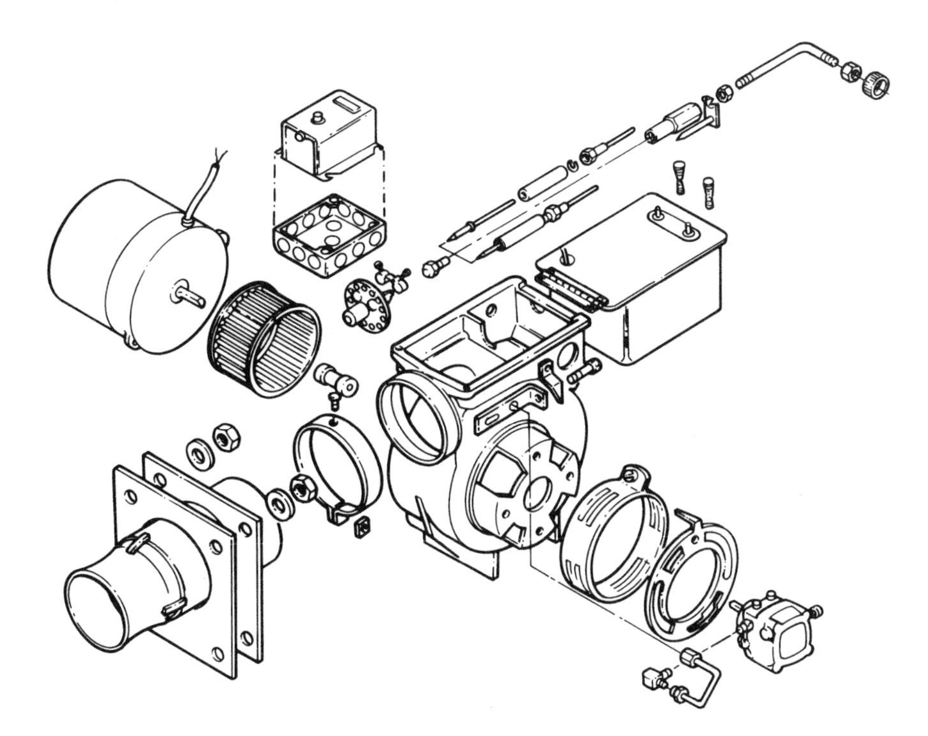

***Fig. 11-21.** A standard oil burner assembly, including part names and letters.*
Courtesy of Blue Ray Company.

Generally, the electric elements are turned On or Off in sequence, one at a time, to supply heat according to demand. The thermostat actuates relays (contactors) as needed. Because electric space heating uses a great deal of power, more capacity is needed, and the service entrance panel might need to be replaced with a larger capacity panel to take care of the much larger load.

ELECTRIC WATER HEATER

The electric water heater is similar to the electric boiler. Water must cover the heating elements whenever the power is on, or the elements will burn out in few minutes. Each month, two to four gallons of water must be drained from the tank. This removes scale and sludge from the tank. Electric water heaters must have their own separate circuit protected by a 40 amp circuit breaker or 40 amp fuses.

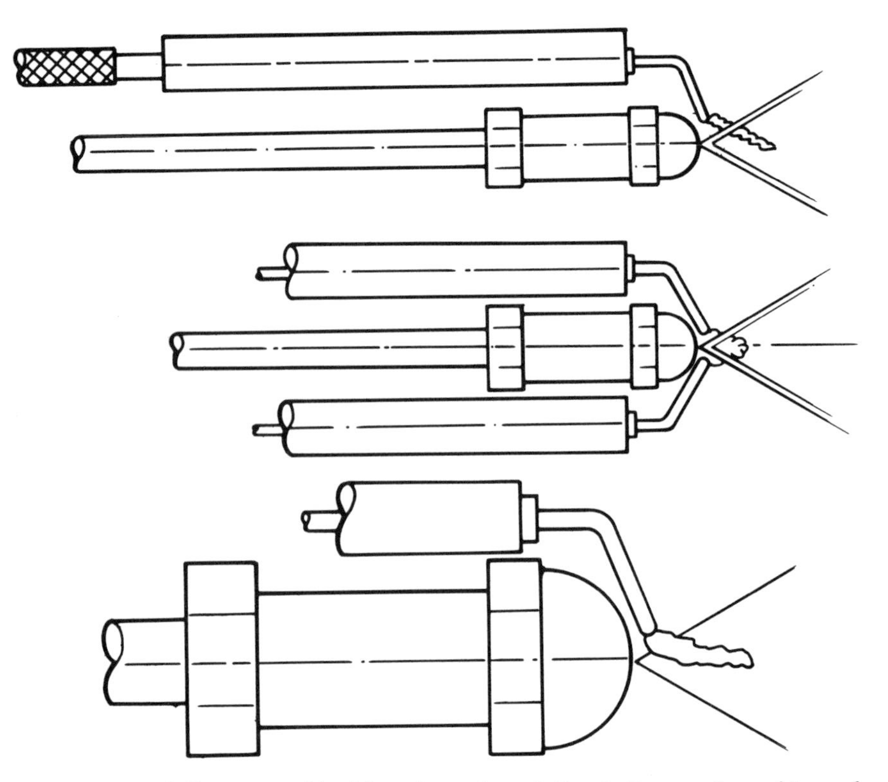

Fig. 11-22. Adjustment of ignition electrodes relative to the nozzle and to each other.

FORCED AIR SYSTEMS

A forced air system, whether for heating or cooling, needs to be balanced to provide equal temperatures in all areas. Proper volume of air is also required in each room.

To start balancing the system, open all of the dampers in the ducts leading off from the main trunk line and open all of the registers in each room. Buy or borrow enough thermometers to place or hang one in the center of each room. Before starting, place the thermometers together in one place for one or two hours, to stabilize the readings. Most thermometers sold in hardware stores will correspond to each other closely enough to use. If one is off a number of degrees, note the amount, so that you can compensate for the difference of that thermometer.

Set the thermostat 5° to 10° *above* room temperature so that the furnace fan will start and warm up the rooms. If it is summer, and air conditioning is included, balancing can be done at that time also. The rooms will cool instead of heat. For cooling, set the thermostat 5° to 10° *below* room temperature at the start of the balancing. Go from room to room and record the temperature of each room in a notebook. Note which rooms increase (or decrease in cooling mode) fastest from the starting temperature. Permit the fan to run long enough to show a definite temperature change, up or down. The room or rooms with the most change should have the duct dampers closed about 1/4 (3/4 open). Allow the fan and burner to operate for another ten minutes after closing the duct dampers, then again check the temperature changes in all the rooms. Make small adjustments of the duct dampers until all of the rooms register approximately equal. You might need to discontinue the balancing if the building becomes too hot or too cool for comfort. I recommend testing when the house is unoccupied, if possible.

12

General Maintenance

NO MATTER WHAT TYPE OF SYSTEM YOU HAVE, REGULAR ANNUAL MAINTENANCE can make your system run more efficiently and save you money. You can also spot problems before they leave you in the cold, such as a worn fan belt or faulty oil gauge.

OIL BURNERS

An oil burner and boiler or furnace should be cleaned once a year for trouble-free operation during the heating season. Start by closing the valve on the oil tank. If it was properly installed, it will include an oil filter, either next to the tank valve or at the fuel oil pump. Spread old newspaper underneath where you are likely to spill some oil. Hold a good-sized container, such as a three-pound coffee can, under the oil filter. Loosen the nut on top of the filter body, but do not unscrew it all of the way out. Jog the body to loosen the neoprene gasket while holding the container underneath. As the filter body drops slightly, oil will run out into the container. Set the container on the floor and hold the filter body as you unscrew the nut on the top of the filter. Pour the oil into the container and also lift the filter cartridge out of the case and set it in the container. Check with the municipality where you live about disposing waste oil and the filter cartridge. Replacements are sold in hardware stores and home centers and include a new gasket.

There are two types of fuel oil pumps on the domestic oil burner, Sundstrand and Webster. Each make differs in design but both operate the same. Each pump has a fine screen strainer under a cover. Many systems have operated for years with this important strainer-trapping sediment because the strainer had never been cleaned. You can get to the strainer by removing its cover. Clean the strainer under hot running water. Replace the strainer using a new cover gasket, available from a heating supply company, see Fig. 6-4.

To get to the nozzle and spark electrode assembly, remove the rear plate of the blast tube (the part that extends into the furnace or boiler). This assembly can then be withdrawn from the blast tube after disconnecting the small oil line from the oil pump and the spark electrodes from the ignition transformer. You can see the connection to the transformer by looking into the blast tube with a flashlight. Have a container and newspapers under the oil pump. Hold one finger over the end of the oil line that was removed from the pump, and drain the oil into the container.

The electrodes are clamped to the oil pipe holding the nozzle. Measure and write down the spacing and position of the electrodes in relation to the nozzle so that they can be replaced in the exact same position. Use two adjustable wrenches to unscrew the nozzle from its holder (see Figs. 11-19 and 11-20). Do not remove the holder from the pipe. The nozzle may have a small strainer screwed into the rear of the nozzle. Remove the strainer if there is one. The nozzle has an insert composed of one or more parts. Older styles of inserts have a screwdriver slot to unscrew the insert. New styles have a hexagon recess and you will have to use an Allen wrench. Hold the nozzle body and the insert parts under hot running water. Be careful not to drop the insert parts down the sink drain as I once did. To gain more force from the sink faucet, attach a garden hose nozzle to the faucet. An old hose connector ring can be used by inserting a piece of sheet metal with a 1/8-inch hole punched in it, add a washer, and screw onto the faucet. Refer to Figs. 3-4, 3-5, and 3-6.

The oil tube that holds the nozzle is usually clean and should not need to be washed. Shine a light into the blast tube as some blast tubes are deteriorated, melted, or broken at the combustion chamber end. A damaged blast tube must be replaced. The dealer who sells that brand of burner should be able to order a replacement for you. All burner installation instructions specify that the firing end of the blast tube *must* be recessed back from the inside of the combustion chamber no less than 1/4 inch. If the tube extends *into* the combustion chamber, the end will melt or deteriorate rapidly. Modern burners have a flame retention head that improves the operation of the burner.

Carefully assemble the nozzle, and insert the internal parts in the correct order (see Fig. 11-19). Tighten firmly in place but do not force. **Note:** If the nozzle has been in use for a number of years, install a new nozzle. After long use, the orifice becomes enlarged sufficiently to pass a larger amount of oil per hour through the nozzle. This will change the oil to air ratio enough to produce a smoky, dirty fire and more soot than normal. A nozzle costs about seven dollars. Take the old nozzle with you for an exact replacement. Specifications are stamped on the nozzle body. For example, .85-80° H means .85 gallons per hour firing rate and an 80° spray angle with a hollow spray pattern.

If the electrodes are in good condition, you can remove the soot with a cloth dipped in mineral spirits. To remove burned on deposits, sand lightly. Rub two pieces of sandpaper or sand cloth together to make them less abrasive then *lightly* sand the porcelain to remove the deposits. Some porcelain tubes have a glossy surface, while others do not. See Fig. 3-6. Remount the electrodes in their holder (as shown in Fig. 3-5) so that their position in relation to the nozzle is correct. The usual measurements are: 7/16 inch *above* the nozzle orifice, the spark gap is 5/32 inch, and the position of the electrode tips is from 1/16 inch to 5/32 inch *ahead* of the nozzle. These critical dimensions vary slightly, depending on the manufacturer's design. In operation, the 5/32-inch spark is blown out into the oil spray and ignites the oil. This design keeps the electrodes out of the oil spray so that they stay clean (see Fig. 11-22).

GAS BURNERS

As gas burners have fewer operating parts, less maintenance is required. Earlier models use a "standing pilot" (the pilot burns all the time). Newer models use an electric ignition, where the main gas valve is lighted by a "glow coil." If the main flame does not light within seven seconds, two more tries are attempted. After the third try, the control locks out. See Fig. 11-12. If there is a standing pilot, it should be cleaned every two years. The burner itself, and all of the other parts should be vacuumed. Turn the gas off and use a flexible hose. Carefully avoid any fragile parts. Vacuum the main burner area and the blower compartment, since they accumulate dirt and lint.

FORCED, WARM-AIR FURNACES

Remove the air filter for inspection and cleaning. Be sure to replace it so that the air flow through the filter is in the same direction as before. If you reverse the filter from its previous position, dirt will be sucked out and blown into the living areas. An arrow on the filter edge shows the direction of air flow. It is a good idea to install a new filter twice a year, especially since the cost is minimal.

The fan and motor generally have oiling ports. Lubricate the blower and motor *sparingly* with nondetergent electric motor oil. Do not use automobile engine oil.

Keep the combustion chamber clean, free of soot and debris. Use a small mirror and trouble light or a flashlight to inspect the inside of the combustion chamber. Also, keep the chimney clean. You can clean it by working a chain up and down while standing on the roof or hire a chimney sweep.

ELECTRIC WATER HEATER

Each month, you must drain two to four gallons of water from the tank to remove scale and sludge from the tank.

HEAT PUMPS

The fan motor and belt are two of the parts you should regularly maintain. The air filters must also be cleaned or replaced regularly. Cleanable filters are popular because they save buying the disposable type. A top quality filter costs about $15 to $20, but will last for ten years or more. If a lesser quality filter is furnished with the unit, and you have bought a top grade type, keep the original to use while the new one is being washed and dried. The filter should fit properly and have no air leaks around the edges. A spacer can be fitted on one side to make a more airtight fit. If possible, inspect the cooling coil occasionally to check for dirt accumulation. A plugged coil can damage an air conditioning compressor, due to frost-up of the coil. There might be an access panel you can remove to inspect the cooling coil.

Check for oiling ports on the fan housing. Most fans now have oil-less bearings that don't need oiling for the life of the motor. If oiling is needed, turn off all electric power before reaching into the fan area.

ELECTRIC MOTORS

It is a good idea to have some knowledge of electric motors that are used in the house and shop. They are familiar to everyone, for example, the furnace fan motor, a bench grinder, or a shop or home vacuum. Nearly every home appliance has a motor, some are accessible, some are not.

Some appliance motors have brushes, such as a vacuum cleaner. Many motors do not, such as a furnace fan and burner motors, and circulating and sump pump motors. These types have two separate electrical windings, a starting winding and a running winding. Because the starting winding must not be energized during the running cycle, it is disconnected from the motor during this cycle. To accomplish this, centrifugal switch is built into the end bell of the motor. When the motor is idle, the switch contacts are closed. When voltage is applied to the motor, the starting winding is connected to the power. As the running winding is always connected to the power, the motor starts to run. This is because the starting winding imparts a torque to start the motor turning. When the motor reaches its designed speed, the starting winding is disconnected from the power because the internal centrifugal switch has opened its contacts. The motor then continues to run on its running winding only.

This motor design does not have brushes to wear down. The contact points of the centrifugal switch might become very dirty. Often these contact points become dirty because of excessive oiling by the owner. I had one instance where I removed a nonworking fan motor from a furnace for repair. As I tipped the motor when removing it, two tablespoons of oil ran from the motor. There should have been no oil leakage, as the bearing wicking holds the necessary oil. I recommend oiling with no more than three drops of electric motor oil in each oil port twice a year. Do not use more than this.

The centrifugal switch is opened and closed by weights on the rotor (rotating part of the motor), spreading and coming together as the motor starts and stops. To clean the switch, remove the motor from its place and take off the end bell where the external wires attach. In rare cases, the switch must be replaced due to wear or burned contacts.

If you have removed the motor, mark the end bells with a center punch, one mark on the end bell and *one* mark on the center section. Do the same with the other end bell, except make *two* marks on the end bell and center section. This designates which end bell goes on which end of the center section for reassembly. Refer to Figs. 4-8, 4-9, and 4-10.

Remove the end bell by first removing the four "through bolts" that hold the two end bells to the center section. After removing the bolts, gently tap all around the circumference at the joining edge with the center the end bell you want to remove. Tap gently so you don't crack the casting. Lay the end bell down next to the motor, being careful of the wires leading to the motor windings attached to the centrifugal switch. The switch points are open when the motor is disassembled. Press on the shiny ring (part of the actuating mechanism) to close the points. Position a business card between the open points, then press on the actuating ring to close the points on the card. Draw the card back and forth to remove the oil and dirt from the points. You might need to use one or two more cards to get the points clean. When the card is clean, the points are clean. The end bell can now be replaced on the motor. Line up the punch marks for a perfect fit. Be sure to avoid pinching any wires while tapping the end bell in place.

13

Related Heating and Cooling Equipment

THIS CHAPTER DESCRIBES EQUIPMENT USED IN CONJUNCTION WITH, OR AS A necessary part of, heating, air conditioning, or heat pump systems. With the exception of thermostats and filters, all of the items discussed are optional. Two such options are humidifiers and dehumidifiers. Humidifiers add moisture to the air and dehumidifiers remove moisture from the air. Properly installed attic fans can be used in place of air conditioning under certain conditions. Electric motors are described and maintenance methods explained. Air filters, essential to both heating and cooling systems, are fully explained. Thermostats are described and illustrated, different types and styles are shown, and the uses of each are explained. Air balancing to achieve equal air distribution throughout the house is detailed, so that all rooms receive equal heating or cooling.

THERMOSTATS

There have been many changes in thermostats in the past thirty years. The original thermostat started the burner on a call for heat and stopped the burner when the thermostat was satisfied. Many newer models can be programmed for various periods of occupancy, lowering or raising the temperature many times during a 24-hour period. Unfortunately, learning how to program complicated thermostat features can be troublesome. Also, the more features included on a thermostat will increase the cost. Usually, two setbacks in a 24-hour period are sufficient. The standard electrical/mechanical thermostat has extra pins, to provide for perhaps three setbacks. This style of thermostat is trouble-free and lasts for many years. Figures 13-1 through 13-5 are representations of various makes and models of thermostats. Many are available in hardware stores and home centers. Although there are detailed directions for installation in each package, note the directions for setting the heat anticipator pointer to match the current draw of the primary control. This setting is 0.45, 0.50, etc.

The location of the thermostat is very important to achieve an even temperature throughout the house. Avoid drafty locations and locations near supply and return registers. An inside wall without strong drafts is ideal. Many people do not realize that the thermostat is an automatic device and maintains the set temperature without any adjustment. Do not change the setting more than two degrees at a time, then allow the temperature to stabilize before making any additional changes.

I have designed a night setback system using two standard thermostats and a twenty-four-hour line voltage timer that works very well. The timer switches control from the day thermostat to the night thermostat at a specified time. I used the present thermostat as the night thermostat and purchased a new thermostat to be used as the day thermostat, then I purchased a line voltage timer. The two new items cost about $60. The old (night) thermostat can be placed in a hallway or other ideal location, or you can even place it beside the new (day) thermostat. The timer should be located in a closet or the basement. See the diagram in Fig. 13-6 for the arrangement and wiring of this setback method. This system is electrical/mechanical and does not involve any electronics, is simple in principle, and is trouble-free. One disadvantage is that there is no carryover, and the timer must be reset after a power outage. This is the same as resetting a standard electric clock, however. See Fig. 13-7.

Thermostats are manufactured for use on various voltages, such as millivolt, 24V, 120V, and 240V. Line voltage thermostats are used with electric resistance heating, and commercially, with suspended unit heaters heating and air conditioning systems. Often, relays can be eliminated when using a line voltage thermostat. Simple On-Off thermostats are used to operate zone valves in hot-water systems (see Fig. 13-17). When a zone valve is opened by the thermostat, an end switch makes contact and energizes a relay, which starts the circulating pump serving all of the zones. Through the pump relay controlled by the end switch units on its own zone valve, *any* one of the zone thermostats is able to start the circulating pump. Conversely, *all* zone valves *must* be closed to stop the pump. This is because *all* of the pairs of leads from *all* of the end

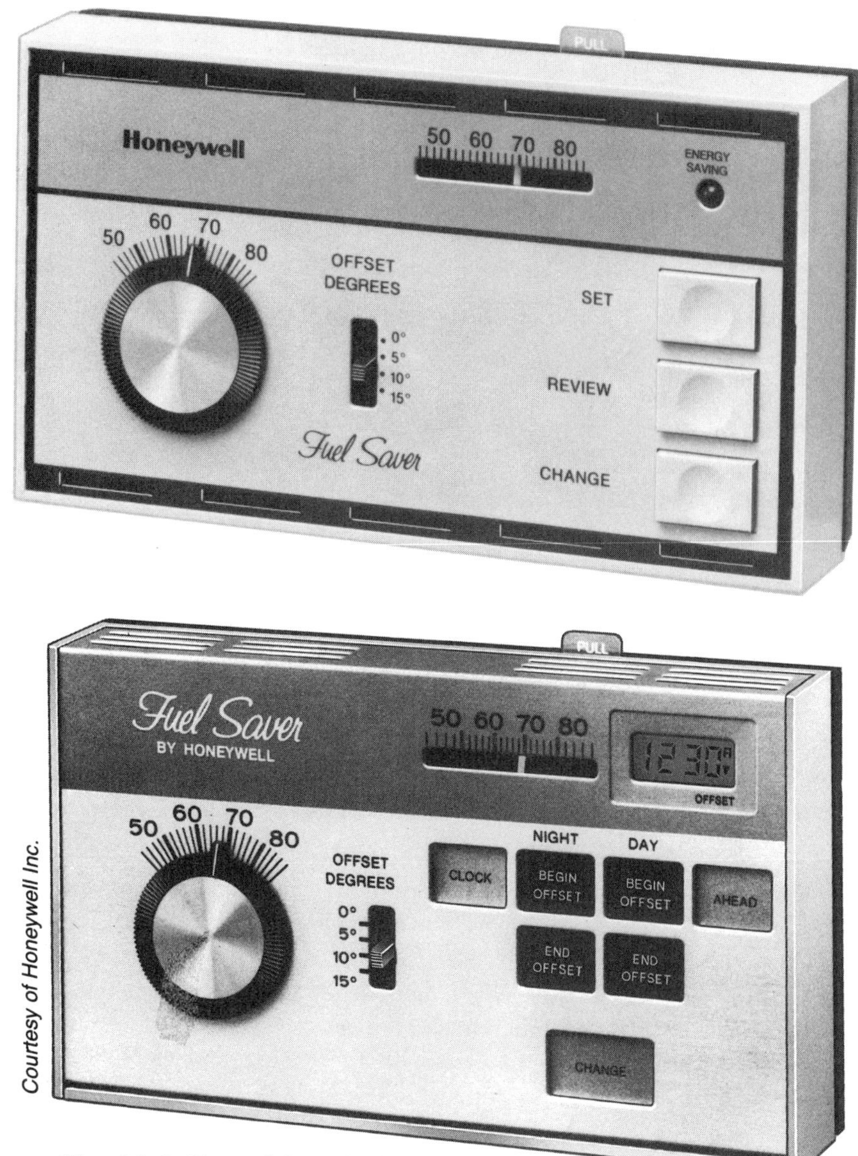

Courtesy of Honeywell Inc.

Fig. 13-1. *Two, night-setback thermostats.*

switches are connected to the same two terminal screws on the circulating pump relay, and are wired parallel.

Thermostats can be used to open and close zone dampers in large, forced air systems or zone valves on large, hot-water systems. At one time, a thermostat was sold to drop the termperature for a selected number of hours during

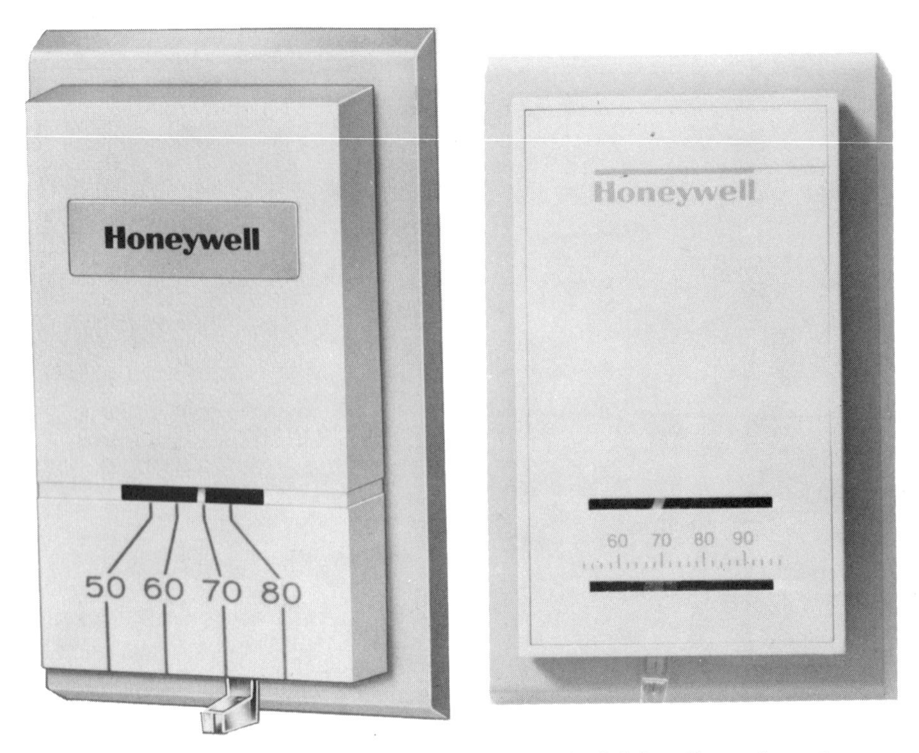

Fig. 13-2. Two thermostat models. The one on the left is a line-voltage thermostat used with electric resistance heating. The one on the right is a heating and cooling thermostat, not a night setback model. It requires a Q634A subbase.

the night by turning a knob on the front of the thermostat, similar to a kitchen timer. At the selected morning hour, the thermostat would reset the temperature back to the daytime setting. This thermostat was reset each evening for the required number of hours. Other thermostats incorporated a small heater inside the case that was energized during the desired setback time by a remote time switch. This increased heat caused the thermostat to sense that the area was at the temperature setting of the thermostat. This small heater was sized so that the drop in temperature was about 8 to 10 degrees below the daytime setting. These simple night setback thermostats are no longer manufactured. They were reasonable in cost and were very reliable.

MILLIVOLTAGE SYSTEMS

During the period when gas heat was becoming available to some areas for the first time, many coal- and oil-fired furnaces and boilers were converted to burn gas. Many older homes still had the older, gravity coal-fired furnaces.

Fig. 13-3. A night set-back and cooling set-up thermostat, and a standard round thermostat mounted on a large coverplate to cover the marks left from the previous thermostat. Courtesy of Honeywell Inc.

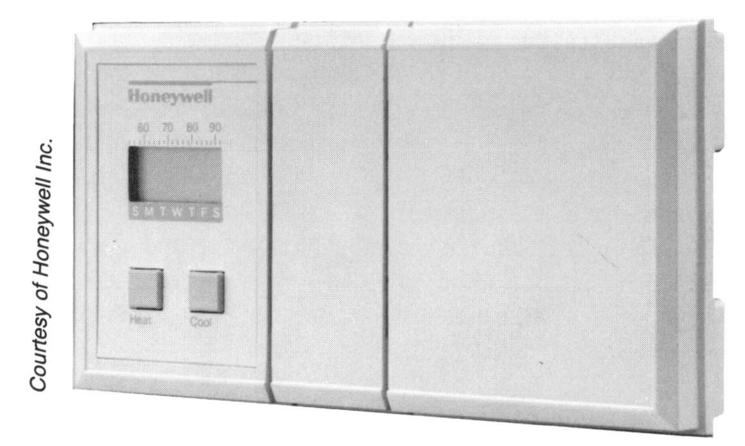

Courtesy of Honeywell Inc.

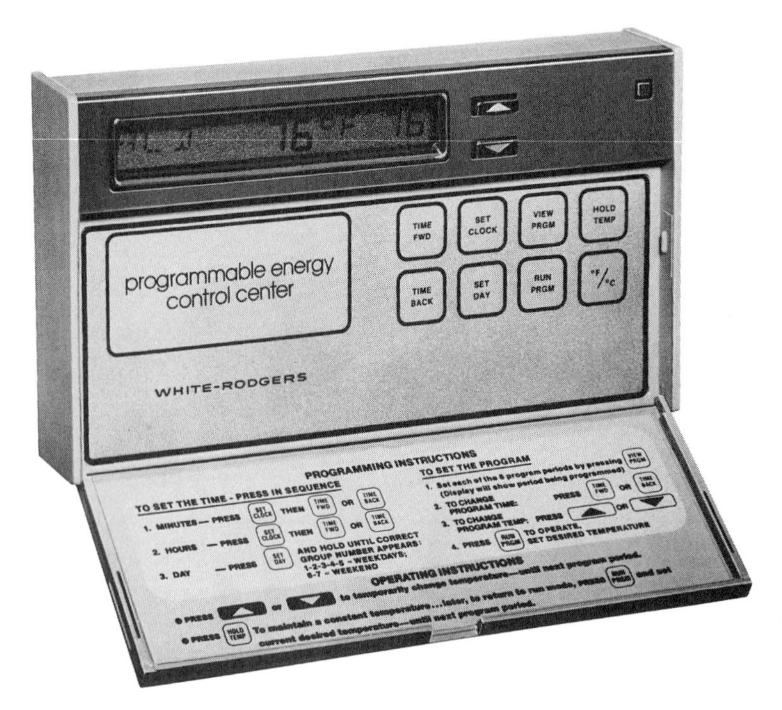

Fig. 13-4. *A programmable thermostat. Courtesy of White-Rodgers Div. and Honeywell Inc.*

These gravity furnaces were readily converted using a conversion type gas burner. These conversion burners could be operated, using one of two systems of wiring. The first system used the standard 24V, which consisted of a 24V solenoid or diaphragm gas valve and a 24V thermostat. The high limit control could be either 24V or 120V. If the 120V limit was used, it would shut off the transformer providing the 24V to the system. If the limit was 24V, only the 24V to the valve and thermostat was shut off.

The second type of wiring system was known as the millivoltage system. This system was powered by a pilot that generated a low voltage of 750 millivolts (3/4 volt). Honeywell, Inc. calls their pilot Powerpile. Automatic diaphragm type gas valves are available for use of the 750 millivolt systems. Thermostats are also available for these systems.

The millivolt system was primarily used on the gravity, hot-air furnace, because automatic heat was provided without any connection to electricity. All power was supplied by the millivolt pilot. This was an advantage when a power outage occurred, and the furnace was still able to supply heat to the house.

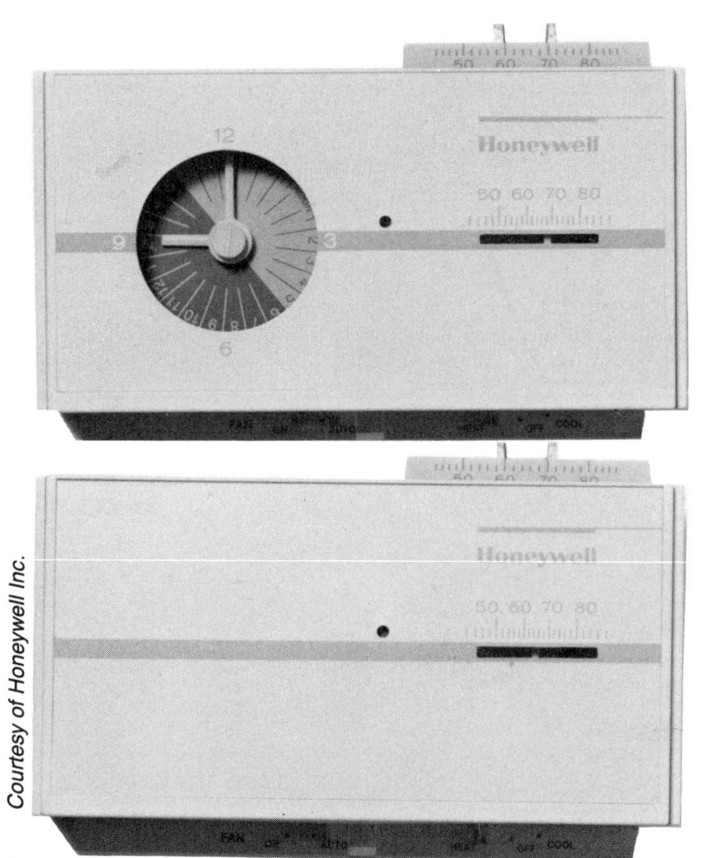

Courtesy of Honeywell Inc.

Fig. 13-5. *Night setback, set-up thermostats (Magicstat).*

HUMIDIFIERS

During the winter season, a furnace humidifier is usually needed to supply the proper amount of moisture into the air circulated throughout the house. The exception is a new house with poured concrete basement walls and floor. These parts of the construction contain a great deal of water. During the first heating season there is usually an excess of moisture in the air and any humidification is not needed. The amount of humidity that is needed is 20 to 30 percent relative humidity (RH). This RH is expressed as the percentage of moisture in the air as opposed to the total moisture that the air can hold at that temperature.

Humidity is provided by installing a humidifier in the supply duct leaving the furnace. Some manufacturers specify that the humidifier be installed in the return air duct. The most common style has a rotating wheel holding a foam pad. At the bottom of the casing is a pan to hold water into which the rotating wheel and pad dip to absorb water. As the wheel rotates, the wet pad gives up its water to the dry air leaving or returning to the furnace, depending on the

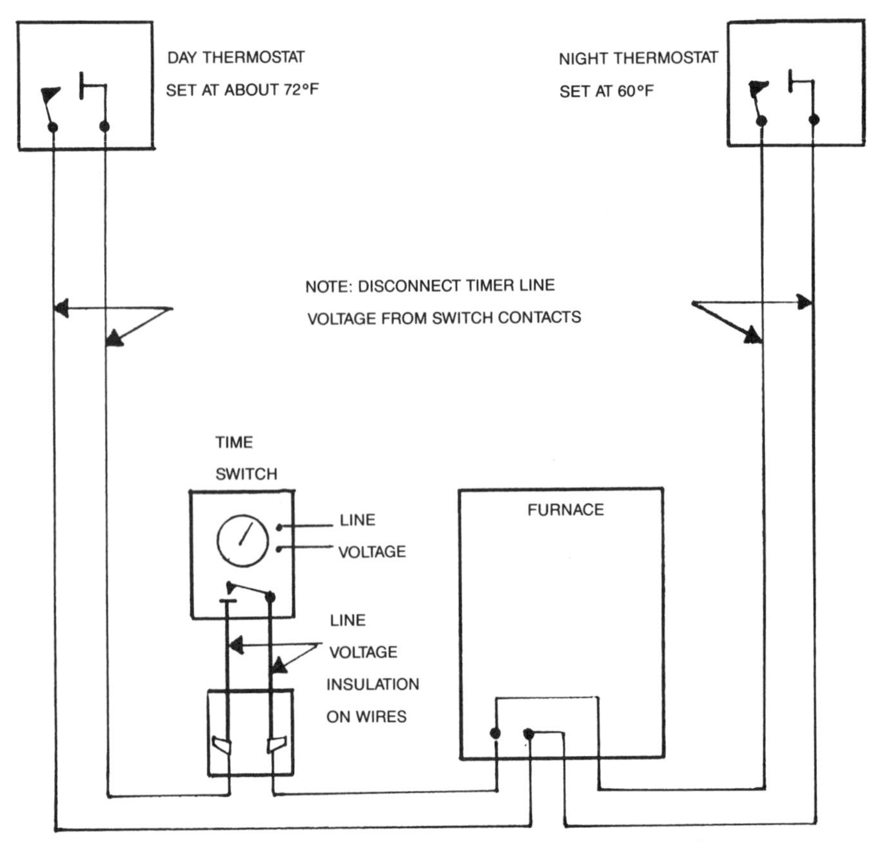

Fig. 13-6. A wiring diagram of a two-thermostat, night setback system.

design of the unit. A small float valve maintains the water level in the pan, and a small, low-voltage (24V) motor rotates the wheel and pad. Many humidifiers using this method have a bypass installed between the supply and return ducts. This bypass has a blank-off plate that *must* be closed in the summer when the air conditioning is used.

Other designs of humidifiers spray heated water vapor into the return duct. These units have been found to leave mineral deposits on draperies and other furnishings however, and are not recommended. Figures 13-8 through 13-16 show various types of humidifiers.

Humidifiers come with complete installation instructions. Some of the tools you will need include: a screwdriver, aircraft snips, an electric drill, and wiring materials. Some humidifiers are equipped with a 120V line voltage, but most use 24V and are supplied with a transformer. The transformer will either be mounted on a square junction box or in a side knockout of an existing pull-chain fixture box in the basement. In most cases, a humidistat is furnished.

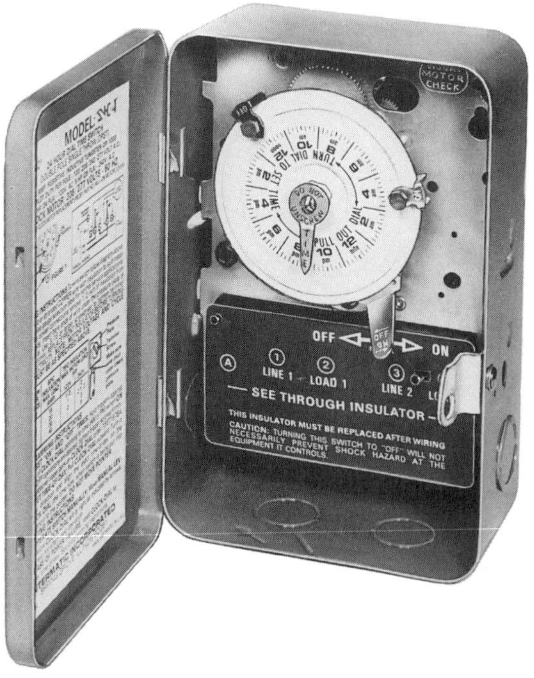

Fig. 13-7. A time switch used with the two-thermostat, night setback system.

Courtesy of Intermatic, Inc.

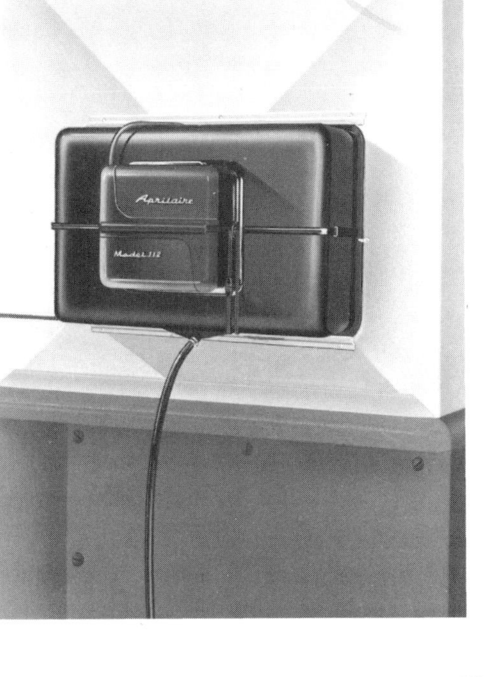

Fig. 13-8. A furnace plenum humidifier. Notice the overflow drain line extend ing from the humidifier.

Courtesy of Research Products Corp.

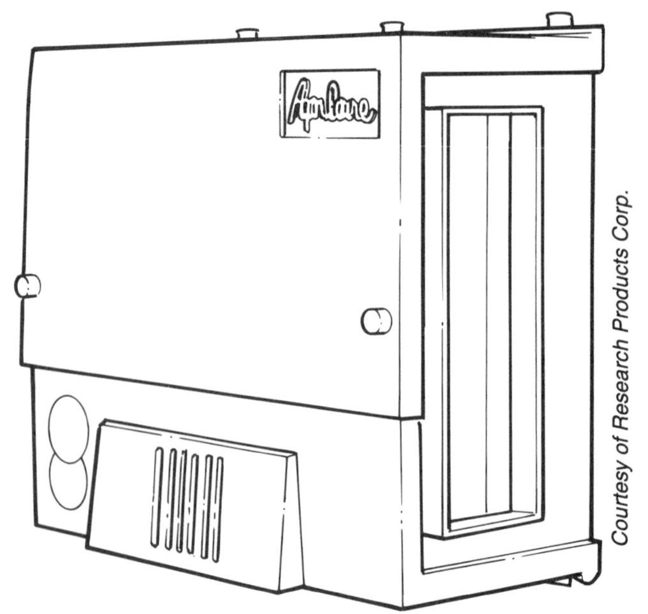

Courtesy of Research Products Corp.

Fig. 13-9. Model 440/445 Bypass type humidifiers.

Courtesy of Research Products Corp.

Fig. 13-10. Model 110/112 humidifiers.

Install this in the return duct *before* the location of the humidifier, *not* after. The humidistat can then monitor the air returning from the living area.

A template is furnished to outline the opening to be cut in the duct for the humidifier. Plastic tubing and a self-piercing saddle valve are furnished to con-

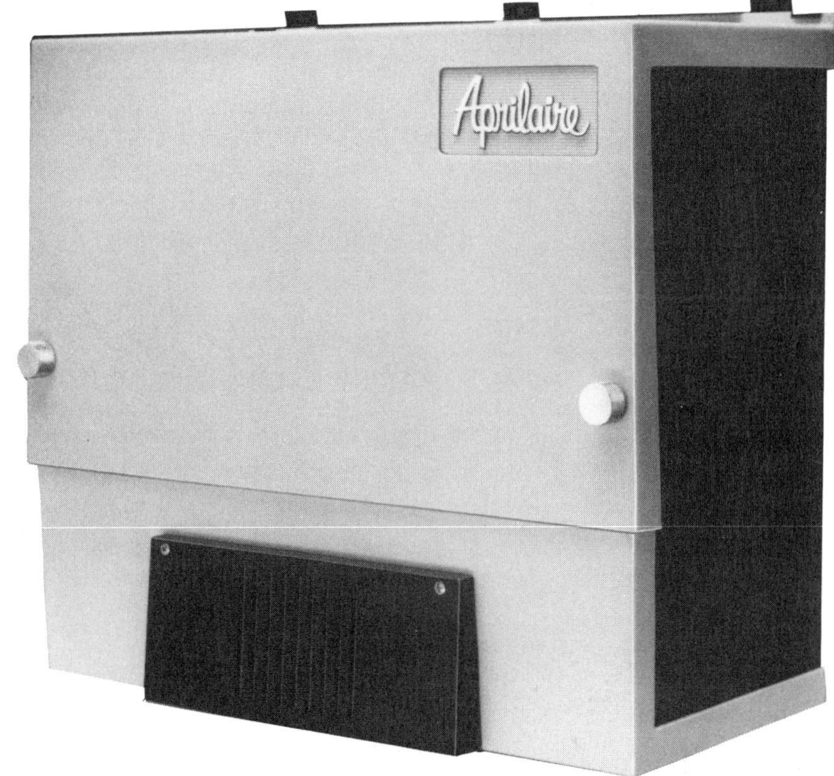

Courtesy of Research Products Corp.

Fig. 13-11. *A bypass type humidifier.*

nect to a cold water line. **Note**: The self-piercing valve can pierce copper tubing, but will not pierce iron pipe. Compression fittings are used to connect the tubing to the valve and the humidifier. If the motor is 24V, connect it to the transformer using an 18-2 thermostat wire, do *not* use doorbell wire. The humidistat also comes with a template to allow its sensing element to extend into the duct far enough to monitor the relative humidity. Too high a setting on the humidistat will cause the windows to sweat.

The humidifier does need some maintenance after the heating season. This is the best time to clean the pan because the scale is softer and easier to remove. Empty the pan and fill it with white vinegar, completely covering the scale. Allow it to soak overnight or longer. After soaking, scrape the scale using a flexible putty knife or similar tool. This is a tedious job, and if the scale cannot be removed, a new pan can usually be purchased from a heating supplier who carries that brand of humidifier. Sometimes a shallow, glass cake pan can be substituted. When using such a pan, make sure that the float of the valve clears the pan sides, operates properly, and maintains the correct water level.

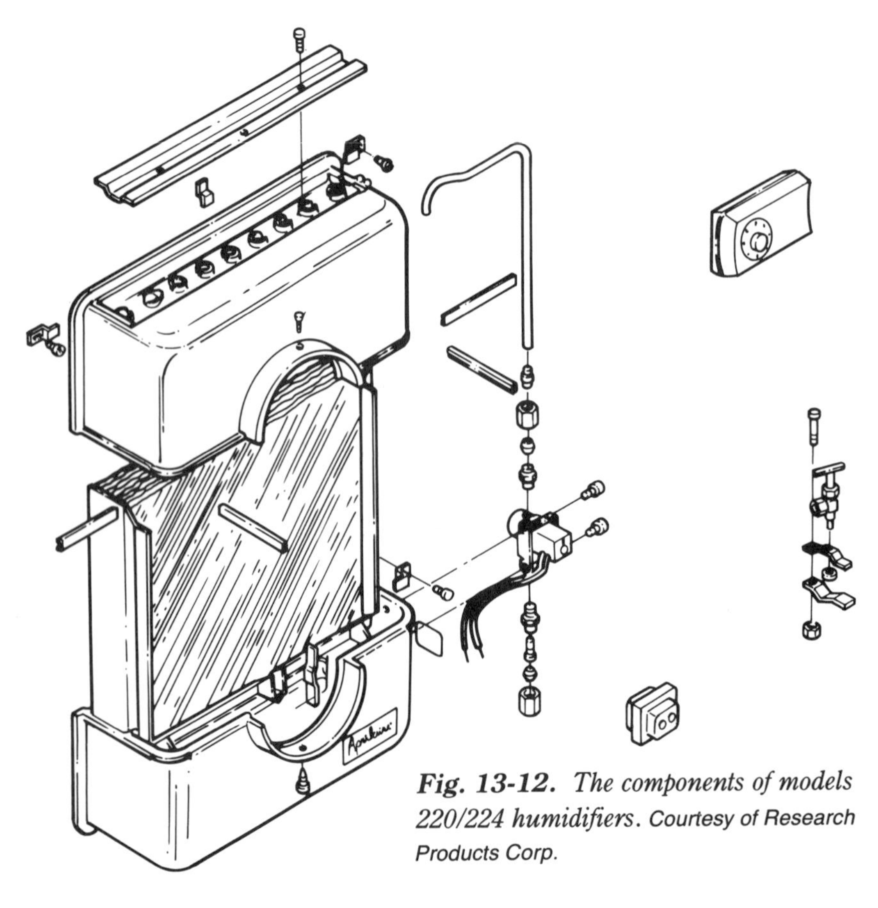

Fig. 13-12. The components of models 220/224 humidifiers. Courtesy of Research Products Corp.

Freestanding (Portable) Humidifiers

If you have hot water or steam heat, you can use a freestanding humidifier to achieve proper humidification. Most portable humidifiers have a water reservoir that must be filled manually, although some provide for a connection to a water supply, similar to a furnace humidifier. Portable humidifiers have a reservoir, a rotating belt or wheel that dips into the water in the reservoir, and a fan that circulates air over the wet belt or wheel and blows it out into the rooms. The output capacity in a twenty-four-hour period varies with the size and manufacturer, anywhere from 2.6 gallons to 15 gallons.

Some manufacturers design humidifiers so that they are attractive pieces of furniture and can be located in hallways or other visible areas. Most cabinets are woodgrain finished rather than real wood. This is because proximity to the water reservoir might cause the wood to warp and split. Both Sears and Toastmaster, as well as other companies, manufacture fine freestanding humidifiers along with furnace humidifiers and single-room humidifiers. All are good quality.

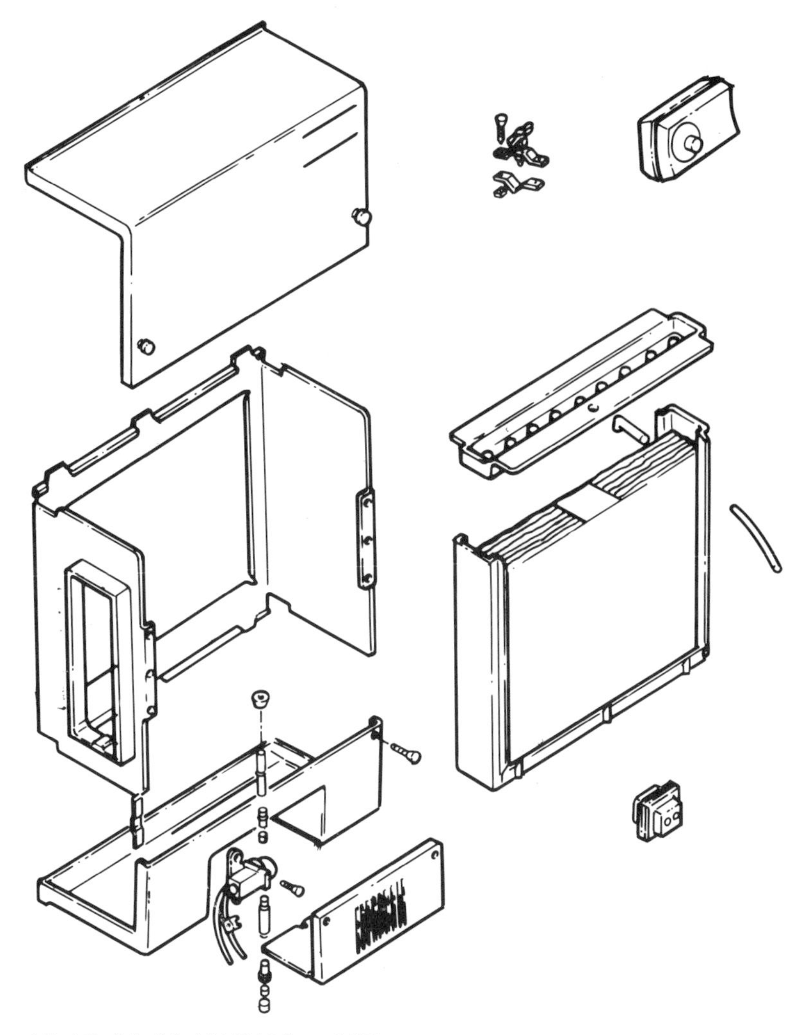

Fig. 13-13. *Model 110/112 humidifiers.* *Courtesy of Research Products Corp.*

Medium-size humidifiers sell for from $119 to $139 and can humidify a medium-size house of 1,500 square feet. Refer to Fig. 13-15.

DEHUMIDIFIERS

Most areas of the country experience high humidity, especially in the spring and summer. Many basements have very high humidity problems, often resulting in leaking basement walls or floors. These conditions cannot be controlled by a dehumidifier and the source of the leaks must be stopped by other means.

Fig. 13-14. A standard humidistat to control the amount of moisture put into a house.

Fig. 13-15. A freestanding, decorative humidifier with a large evaporative capacity of 12 gallons every 24 hours. It has a rotating foam belt that picks up water from a reservoir.

The so-called "normal" high humidity basement, often called a "damp basement," can be readily controlled by a modern dehumidifier that is properly sized for the area of the basement. An adjustable sensor is included to shut the dehumidifier off when the humidity has been reduced to an acceptable level. Some sensors have a constant On setting, while others have no control and will run continuously when they are plugged in. A reservoir is provided to catch the condensate (moisture) removed from the basement air. Do not reuse this condensate water; it might contain mold spores and other contaminants. If there is

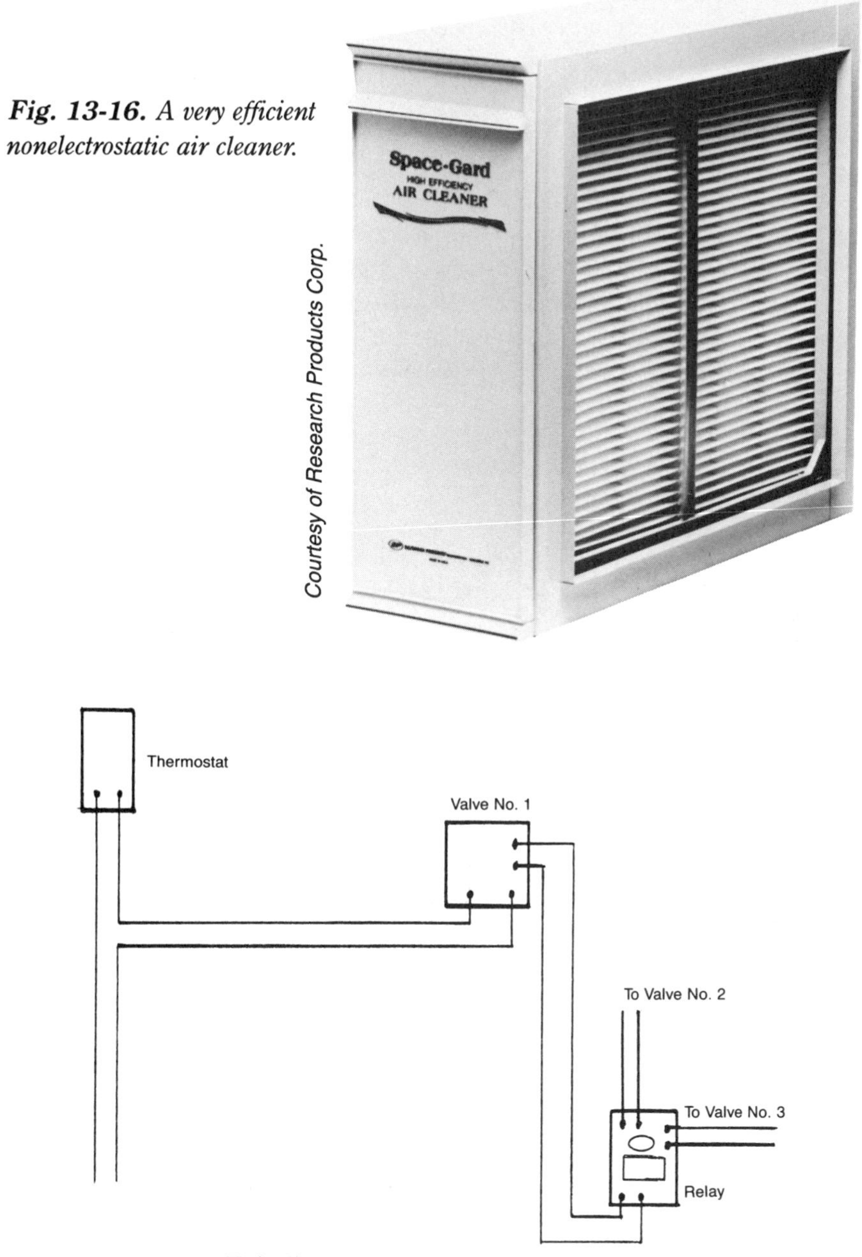

Fig. 13-16. *A very efficient nonelectrostatic air cleaner.*

Fig. 13-17. *This diagram shows how to wire zone valves with end switches incorporated into the zone valve motor.*

a convenient floor drain nearby, you can set the dehumidifier over the drain after the reservoir has been removed. Then the condensate can drain directly into the sewer. This eliminates the task of checking and emptying the reservoir. Some dehumidifiers have a garden hose connection on the drain spout, and a hose can be connected to a drain location.

The dehumidifier should be plugged into an outlet that is always On, such as a wall outlet, to provide good control of humidity. The cooling coil and condenser coil need periodic brushing to maintain efficient operation and save electricity. Unplug the unit before brushing the coils. **Note**: Always make sure any automatically operated equipment is disconnected or locked in the Off position before attempting to work on the equipment. Many accidents have happened because people do not follow this advice.

AIR FILTERS

Air filters are very important to the operation of a forced air system, whether it is a furnace or a straight air conditioner. Air filters are especially important if central air conditioning is part of the system. The air conditioning coil is designed with closely spaced fins that are pressed on the actual refrigerant tubes. These fins collect a great amount of dust and dirt. If the incoming air is not filtered, the finned coils will quickly plug up. Many air conditioning coils are very difficult to clean, because of limited access to the face of the finned coil. You might even have to cut an access opening in the plenum sheet metal in order to clean the coil. Casings furnished with cooling coils might have an access opening already provided. Filters are also helpful in keeping furnishings clean and dust free. Many permanent or washable filters also can catch pollen and other allergens.

Filters for home heating and air conditioning come in many sizes. One type permits the homeowner to cut the filter to the required size. This makes it convenient when the system requires an odd size filter that is not available in stores. All filters used in home equipment must be no more than one inch thick, because the extra thickness will restrict the airflow. The furnace blower (fan) is not able to overcome the extra static pressure caused by the two-inch filter and this will cause problems.

It is important that air filters fit the space provided for them. A loose fit allows air to be drawn *around* the filter instead of *through* the filter. The filter rack can sometimes be made smaller by fitting metal strips on the top and one side. These strips can be fabricated at any sheet metal shop for a small cost.

When removing washable filters for cleaning, mark an arrow on the filter edge to indicate the direction of airflow through the filter. This ensures that the filter is replaced with the air flowing through in the original direction. This is useful when the filter is removed for inspection only, and will not be cleaned. If the filter is reinstalled backwards, any residual dust or dirt will be pulled from the filter and blown back into the rooms. This defeats the purpose of the filter.

Electronic Air Filters

Electronic air filters were developed to more perfectly clean the air than the disposable filter. Electronic air filters can remove up to 95 percent of the particles in the air, including tobacco smoke and plant pollen, among others.

This filter operates by ionizing particles in the air with an electrical charge from wires strung across the air duct. Beyond the wires are a series of plates with the opposite charge of electricity. Opposite charges attract, so the charged particles are attracted to the plates and are caught on them. The plate assembly is removed each month, and the plates are washed to remove the particles.

The filter is installed at the return air connection to the furnace. Any changes to the return air connection can be done by a sheet metal shop. The unit operates on 120/240V. The cost of the filter is about $300 to $500 depending on the size and brand, plus installation. Many find that a cleaner house and no pollen to breathe is well worth the investment.

ATTIC FANS

During the warm months, when nights are cool, the attic exhaust fan serves a useful and economical purpose. At dusk, windows are opened an inch or two, the attic fan is turned on, and the cool night air is drawn in through the slightly opened windows. The hot air from the living area and the attic is then exhausted through the attic ventilating louvers. A properly sized propeller fan can cool down a house to the outdoor temperature in about an hour. Some fans include a thermostat or timer, because many times the house becomes cooler than desired.

Fans are available in various diameters, and the size you need depends on the square foot area of the house. The handyperson can fabricate a suitable fan support. A motor might be available from the home workshop, or can be purchased second hand. The best arrangement is to use a short shaft and mount two bearings on an A-frame of wood or metal. Install the shaft in the bearings and mount the fan blade on one end and a pulley on the other. Mount the motor on the A-frame side. The pulley for the motor should be smaller than the fan pulley to reduce the speed to less than the motor speed. Wire the motor to a convenient switch. If the fan is to be mounted over the attic access, you must build an enclosure around the fan and motor. This arrangement only exhausts air from the living area, *not the attic*. A better arrangement is to mount the fan to blow directly out through the attic louvers. This method also requires an enclosure so that the fan can only blow out through the louvers. This method also exhausts air from the attic, providing increased circulation and cooling. **Note:** If an enclosure *is* built around the fan assembly sitting directly over the attic access opening, a door *must* be built so that access to the attic area is easily available. This is very important in case quick access must be gained to the attic because of a fire or other emergency.

The smallest useful fan diameter is 20 inches, with a 1/4 horsepower motor. Larger fans, up to forty-two inches diameter, use a 1/2 horsepower motor. A professional installation mounts the fan in the attic on rubber isolation mounts for quiet operation. A gravity-operated damper is mounted horizontally in a ceiling opening under the fan location. The damper is the louver type and is opened by suction of the fan. When the fan stops, the blades close by gravity.

The reason that the attic fan is so desirable is that it exhausts the very hot air from the attic to the outside. Even though ceiling insulation is installed, using the attic fan keeps the living areas cool. I have used both an attic fan and central air conditioning in the same house. Many times the attic fan was used instead of the central air unit. This was an economical and very satisfactory alternative to central air conditioning. Either the attic fan or the central air conditioning can be used, but not both together, depending on the outside temperature. For hot, daytime weather, use the central air, for cool evenings, use the attic fan.

Troubleshooting Gas Burners

PROBLEM	SOLUTION
Burner off, no heat	Pilot light is not burning. Relight pilot by following the directions on the burner nameplate. Fuse blown or circuit breaker tripped. Reset breaker or replace fuse. Thermostat set below room temperature. Thermostat contact points are dirty (older models). Draw business card between closed points. Some conversion burners have a manual switch (Baso brand) that must be reset by pushing down on the knob on top of the case, *after* lighting the pilot.
Burner off, pilot burning	Check to determine if the Baso switch has been reset. On newer burners, check to determine if the knob on the automatic valve is turned to the On position. Turn knob to On position. Burner should light.
Burner off, pilot off	Check if vent from utility gas regulator (inside) is open. Vent is on outside of house wall. Remove vent screen and clean screen and vent.
Burner off, pilot burning	Thermocouple could be defective. Remove and take to hardware store to purchase identical thermocouple. Pilot flame does not impinge on thermocouple tip. Flame might be too small to heat thermocouple tip. Remove pilot and clean.
Burner on, pilot on. Not enough heat. Burner goes on and off	Filters might be very dirty or clogged. Replace filters. Fan belt could be loose and fan does not turn fast enough to remove heat from furnace. Belt could be broken or off the pulleys. Fan or motor pulleys could be loose on shafts and fan cannot run. Use an Allen wrench to tighten pulleys.

continued

PROBLEM	SOLUTION
Burner will not light. Pilot burner is on	Check furnace transformer, 120V to 24V for voltage. Transformer might be burned out. Check voltage at furnace. Fuse or breaker could have blown or tripped.
Conversion burner in gravity hot-air furnace will not come on	Millivoltage pilot generator could be defective. Remove and replace. Check thermostat. Clean open points, using a business card drawn between points pressed together. Thermostat might be old and need to be replaced. Buy a two-wire low voltage thermostat with mercury bulb contacts.

Troubleshooting Gas Furnaces

PROBLEM	SOLUTION
No heat	Pilot out. Relight by following directions on the furnace instruction plate. No power. Check fuses or circuit breaker.
Fan belt broken.	Replace fan belt; available at hardware stores.
Fan motor will not run.	If motor hums, internal switch contacts are dirty. Remove motor end bell and clean points.
No gas pressure.	If there is a pressure regulator on the gas line inside the house, outside vent from it may be plugged. Remove cap on outside end and clean screen if used.
Fan or limit control defective.	Check and replace if necessary.
Squirrel cage blades full of dirt.	Remove complete fan and clean blades.
Gas valve will not open but pilot burning.	Main valve defective. No power; check fuses or circuit breaker.
Thermocouple defective; tighten end at main valve just snugly.	Replace thermocouple if above method fails. Available at hardware stores.
Pilot tip not hot enough. Dirty thermostat contacts. Draw business card between points. Late model thermostats have sealed mercury contacts, cleaning is not needed on these.	Flame may not be hitting thermocouple properly. Flame may be too small; increase size of flame using adjusting screw on main valve.
Broken wire in thermostat wiring cable.	Short out terminals on gas valve marked "thermostat" or "thermo." Burner will start.
Furnace noisy	Fan bearings are worn. Replace bearings. Available at mill supply company and some hardware stores.
Burner pops on start	Too much air. Close air shutter on gas burner tubes.

Troubleshooting Oil Furnaces

PROBLEM	SOLUTION
No heat	No power. Check fuses or circuit breakers.
Burner will not start	No oil. Call oil supplier for emergency delivery. Stack relay will have locked out. Turn furnace switch off, reset stack relay. Turn furnace switch on, if burner does not start, check limit switch by shorting switch terminals to bypass switch. Also the stack relay may be defective. Burner oil pump may have to be primed. There will be a pipe plug opposite where the oil line attaches to the pump. Unscrew this plug ONLY partially. First air will come out and then oil will seep out. When the seepage runs for a few seconds, tighten the plug and restart the burner. Stack relay defective. Replace relay. Limit control defective. Replace limit control.
Flame established, but has acrid odor and is white, not orange	Excess of air or the oil nozzle is partially plugged. Close air shutter slightly, or remove complete blast tube assembly (nozzle, oil tube, ignitors, and high voltage wires). Remove nozzle, disassemble nozzle (three or four parts), wash in very hot water and reassemble. Replace in nozzle holder, reinstall assembly and make the necessary connections. Restart burner.
Burner runs but no flame. Has plenty of oil in tank.	Oil tank ran empty, but was refilled. Pump has lost its prime. Prime pump. Burner oil pump may have to be primed. There will be a pipe plug opposite where the oil line attaches to the pump. Unscrew this plug ONLY partially. First air will come out and then oil will seep out. When the seepage runs for a few seconds, tighten the plug and restart the burner. Burner nozzle plugged. Remove complete blast tube

continued

PROBLEM	SOLUTION
	assembly (nozzle, oil tube, ignitors, and high voltage wires). Remove nozzle, disassemble nozzle (three or four parts), wash in very hot water and reassemble. Replace in nozzle holder, reinstall assembly and make the necessary connections. Restart burner. Oil filter plugged. Replacement available at hardware stores. Pump strainer plugged. Clean strainer. The pump strainer is internal under a round or square cover. The strainer wil be adjacent to the oil line connection to the pump. These strainer covers take a special gasket available at wholesale heating suppliers or pump repair companies. Heavy bond letter paper may work in an emergency. NOTE: These strainers are seldom cleaned and are a major source of trouble if not cleaned. Clean in very hot water.
Fan does not run	Fan control defective. Replace control. Check the fan control by shorting across the fan control terminals. Use a piece of insulated #14 or #12 wire. Touch the bared ends of this jumper to the terminals. The fan will start. Remember that for the control to act on its own, the furnace bonnet must be quite hot (about 130°). When you jump the control, you are bypassing the control, and if the fan motor is good, it will start even though the furnace is cold. This only proves that the fan and motor are in operating condition.
Furnace noisy	Fan bearings are worn. Replace bearings. Available at mill supply and some hardware stores.

Troubleshooting Electric Resistance Heating

PROBLEM	SOLUTION
No heat, fan does not run	Fuses or circuit breaker could have blown or tripped. Replace fuses or reset breaker. Sail switch might be defective. This switch senses air movement and makes the circuit on airflow, and breaks the circuit on *no* airflow. The switch has a 3″ × 4″ flapper plate that can break off and cause the switch to be inoperative.
No hot water (boiler)	Boiler might be off on high limit. Limit might have to be reset manually. Fuse blown or breaker tripped. Replace fuse or reset breaker.
No heat	Boiler heating elements might be burned out. Replace elements. Turn power Off. After marking each wire to show to what terminal it belongs to, disconnect the wires from each one. Test across the two terminal screws with an ohmmeter. An infinity reading indicates a good element. A Zero reading indicates a bad element. Replace the element.

Troubleshooting Forced, Warm-Air Furnaces

PROBLEM	SOLUTION
No heat	Turn furnace switch Off. Check fuses or circuit breakers, replace fuses or reset breakers. Check thermostat, it could be set below room temperature.
Limit switch is in the open position, but burner will not start	Check if furnace has been On position, recently (is warm). Check if limit switch is in open position. Some limit switches must be reset manually. Fan belt could be broken or off of the pulleys. Belt could also be very loose. Filters might be dirty or plugged, causing the limit switch to shut the burner off.

Troubleshooting Oil Burners

PROBLEM	SOLUTION
Burner locked out	Check for blown fuses or tripped circuit breakers. Check if safety relay has tripped and reset once. If burner lights and runs, this was probably an isolated instance. If burner starts, but does not light, check the oil tank, it might be empty. Use a stick to check the oil level. Do not depend on the tank gauge. NOTE: Tank might have run empty, and *was* filled, but burner oil pump might need to be primed. Loosen the pipe plug on the pump opposite the oil line connection from the tank. **Do Not Remove**! Loosen just enough to allow air to seep out around the threads until oil appears. Then put the plug tightly back in place. Reset the safety relay. The burner should light and run properly.
Burner locks out again. Motor hums, but does not start.	Turn off power. Pump failure or freeze-up. Open burner blast tube at burner rear end. Check if fan cage can be turned (some resistance will be felt). If fan cannot be rotated, pump bearing or motor bearings are frozen. Loosen the coupling from the pump and fan. Coupling is fastened with an Allen setscrew, tightened against shafts. As motor and pump are separated, each screw can be checked for tightness separately. Remove pump or motor for repairs. Many wholesalers have pumps and motors for exchange or sale. Be sure to trade "like-for-like," rotation, shaft size, and mounting bracket design.

Troubleshooting Heat Pumps

Many times, glitches occur in automatic equipment that have no logical explanation. These glitches are voltage fluctuations and momentary blackouts that can occur when a utility line fails and the system automatically transfers the power to another circuit. This results in a period of two or three seconds without power, and air conditioning equipment can trip out when called upon to restart after being in the running mode just a few seconds earlier. If another heavy load is put on the line at the exact time that the equipment starts, or any other minor malfunction of equipment happens, glitches can occur. These glitches *are* frustrating and *do* occur without any reason being found.

PROBLEM	SOLUTION
Outdoor coil frosts up	Unit has not gone into the defrost cycle. Check wiring.
Not enough heat in building	Outdoor temperature might be below 35°F. Supplementary heat might not have come on. Check fuses or circuit breakers for furnace heating elements. Check fan motor circuit, electric heating elements will not be energized if fan does not operate. Check auxiliary heating equipment.
System will not switch over to heating mode	Check the wiring and relays. Check for blown fuses or tripped circuit breakers. Check for defective relays and loose or defective wiring connections.
System will not switch over to cooling mode	Check wiring and operation of relays. Check for blown fuses or tripped circuit breakers. Check for defective relays and loose wiring connections. Check for breaks in wiring.
No heat or no cooling	Usually the main fuses or main breaker for the heat pump system has blown or tripped. At times, this is because of low voltage on the utility lines. Check the voltage if you have a voltmeter. Call the utility company and have them check the voltage. When the unit tries to start on low voltage, it will draw more amperes and trip the breaker or blow the fuses.

continued

PROBLEM	SOLUTION
Low voltage	Can be caused by undercapacity of the original service entrance panel, or loose and corroded terminals in the panel. Wiring that is undersized is feeding the condensing unit from the service entrance panel. Have the local electrical inspector approve tapping off the *load* side of the utility meter on the outside of the building and running a line directly to the condensing unit fused disconnect.
Power failure	Call the utility if no voltage is found on the premises.
Unit makes a loud clatter when starting	Turn off the power to the unit and remove the panel cover. Inspect the contactor for burned points.
Points on contactor are badly burned	Call service technician for a complete check of the system.
Does not heat or cool properly	System might be low on refrigerant. Check liquid line sight glass. If bubbles show, call technician experienced on refrigeration and heat pump systems.

Troubleshooting Air Conditioners

Many times, glitches occur in automatic equipment that have no logical explanation. These glitches are voltage fluctuations and momentary blackouts that can occur when a utility line fails and the system automatically transfers the power to another circuit. This results in a period of two or three seconds without power, and air conditioning equipment can trip out when called upon to restart after being in the running mode just a few seconds earlier. If another heavy load is put on the line at the exact time that the equipment starts, or any other minor malfunction of the equipment happens, glitches will likely happen. These glitches *are* frustrating and do occur without any logical explanation.

PROBLEM	SOLUTION
No cooling, nothing runs	Check fuses or circuit breakers. Reset breaker or replace fuses.
Unit starts, but stops	Reset again.
Unit starts, but stops again	Call service technician.
Little cooling	Check for clogged filters. Check for loose fan belt. Check sight glass. If bubbles show, system is low on refrigerant. Check for leaks using a soap solution, or a halide torch.
Cooling coil frosts up	Not enough air passing over the cooling coil. Check for closed registers or closed dampers in duct lines. Check that any humidifier bypass is closed.
Compressor stops and starts frequently (cycles). Cooling coil frosts up	Low on refrigerant. Check sight glass. Check on proper operation of fan. Fan might have loose belt. Check air filters. Check for leaks.
Compressor noisy on start-up	Turn off power and remove cover from panel. Check wiring and relays. Check to see if points on contactor are burned. Replace contactor points, or call service technician to replace them.

continued

PROBLEM	SOLUTION
Compressor hums, but does not start. Trips overload or blows fuses	Contact points might be burned. Compressor might be burned out. See solution above.
Compressor seems to be laboring	Check that leaves and grass are not blocking fins on unit. Check that fan runs fast enough. Check that voltage is within plus or minus 10 percent of 240V.
Water coming from cooling coil	Drain line is plugged. Disconnect line to clean drain pan and drain line.
Drain line pump does not run, or runs continuously	Check float system of pump. Float could have a hole in it. Buy a new float where you purchased the pump. Check the fuse for this circuit. It might be on the furnace circuit.

Troubleshooting Wood Stoves

PROBLEM	SOLUTION
White spots on stove	Furnace cement can bloom (turn white). Thoroughly brush any area that turns white with a wire brush to remove the cement so that the spots do not reappear. Don't paint over them without first brushing or they will probably reappear later.
Dirty glass	Burning wood, particularly during slow firing, causes creosote to coat the windows with a blackish film. Scrape off with a putty knife, clean with a chemical solvent, or burn off during high firing.
Paint looks bleached look at both sides	If paint pigment is burning out, you are probably over-firing. Repaint and rein in on your heating ambitions.
Shaker grate jams	Rocker grates can jam when fuel gets caught between the rocker grate and the fixed center grate section. If one section locks up during shaking, do not try to free it, you could break the grate. Wait about an hour and the fuel causing the jam will burn up and the grate will be freed up. When shaking, the grate can rotate 90 degrees. For ordinary shaking during an ongoing fire, however, the rotating grates should only be rotated 45 degrees. A full 90-degree shake will dump the coal into the ashpan and might even cause the grate to jam.
Low draft	Low draft can be caused by shorter burn times, low heat output, inability to maintain a fire, and smoking. Your chimney or stove pipe could be blocked up with creosote or some other substance. Inspect the pipe or chimney

continued

PROBLEM	SOLUTION
	periodically to avoid this. A connection between two sections or stove pipe might have come loose. Seal the connections with anchor screws. Isolate leaking joints using a lit cigarette (its smoke will be drawn up). The pipe installation might have too many turns or run horizontally too long. Turns create resistance to the smoke's path. Avoid more than two, 90-degree turns in two steps, with two 45-degree elbows to minimize the draft resistance. To increase the draft despite too many turns, consider extending the height of the chimney or pipe extending into the chimney. **Short Stack**. Short stacks can cause bad drafts. The minimum height should be at least 18 feet. The chimney should be at least three feet above the roof line and two feet higher than anything else within a 10-foot radius horizontally. **Oversized Chimney**. If the area inside the chimney is too large, the smoke will disperse and rise slowly, limiting the draft. For a 6″ flue-size stove, the chimney area should be 30 to 50 square inches. For an 8″ stove, the chimney area should be 50 to 90 square inches. If the chimney is too large, reduce the interior size by adding chimney liner or help the draft by extending the pipe from your stove well up into the chimney. **More than one heating device is vented into the stack**. Only one heating device should be used per flue. The draft might be bypassing your stove and drawing through the other device. **Poorly sealed installation and positive connections**. Improperly sealed connection from stove to stack can be permitting the draft

continued

PROBLEM	SOLUTION
	to skip the stove and draw through other openings, particularly if vented into a fireplace. Make sure all connections are sealed tightly. Look for leaks in the stack itself. If it is a masonry chimney, look for poorly fitting ashdoor in the basement or on an outside wall. Place a lit cigarette near connections in your chimney or block-off plate when the stove is operating. If there is a leak, the smoke will be drawn up. **Undersized chimney**. The chimney must have an interior square area at least equal to the stove's flue pipe. **Tightly sealed house**. A stove depends on a house's natural leakage to replace the air it burns and sends up the chimney. If your house is especially well-sealed it might prevent this leakage. Test by cracking open a door or window close by. If it makes a difference, install an outside air intake to bring in combustion air for the stove to burn from the outside of the house.
Can't get long burn time	Besides a poor draft, airtightness is also required for long burns. Over time, cemented seams can deteriorate. Some cement might also loosen when the stove is in transit. If your stove is running too fast, examine all internal firebox seams for evidence of leaking. If leaky seams are found, use a screwdriver to first clean the seams out and apply fresh cement. Examine the fit of your doors. The door gaskets can compress. Periodically replace them. Use a small amount of furnace cement to hold them in their channels. Test the fit by closing and latching the door on a slip of paper. If the paper pulls out without ripping, the gasket isn't tight enough. Adjust by packing more stove cement under the gasketing.

continued

PROBLEM	SOLUTION
Can't maintain a coal fire	If stove burns well but can't maintain a coal fire, this indicates that there is a draft problem or you need at least a 30-degree difference between the indoor and outdoor temperature to maintain a sufficient draft for coal burning. It is difficult to maintain a coal fire on warm days, use wood early and late in the season and burn coal only when it's cold outside.
Not getting enough heat	The area could be poorly insulated or drafty and might not retain heat well. You might have a draft problem that is limiting your maximum firing level or the stove might be undersized for the job.
New stove's surface smoking	The stove paint and furnace cement "cure" and let off smoke during the first few burns.
Stove or stove area smoking	Determine exactly where the smoke is coming from. If it is coming out of the front and side of the stove in repeated puffs, then it's a downdraft problem. If smoke is consistently coming out of one part of the stove, one of the seams or the doors might be leaking (see other entries for these problems). Apply fresh stove cement to the seams to patch leaks. If the smoke comes from the pipe area, the pipe connection or block-off plate is probably not sealed or has come apart. Reseal or reconnect. When opening the front and/or side doors, the bypass gate must be opened first. Otherwise, the stove will smoke. If the bypass gate is open and the stove is still smoking, then it's probably a draft problem.
Downdraft	Frequent downdraft problems could mean you need to increase the height of the pipe extending from the stove or increase the chimney height. Check for nearby trees or other

continued

PROBLEM	SOLUTION
	obstructions that might interfere with draft and remove any obstructions. Check using a regulating device such as a Vacu-Stack downdraft preventer.
Doors not sealed properly.	Using the removable door handles, rotate the locks to tighten or loosen the latches as necessary. This will keep the door well sealed. Reset the catch on the threaded stem with an Allen wrench. The gasketing might also lose its original shape and prevent the door from closing tightly and you should replace it from local hardware store.
Can't get sufficient wood into firebox	Try using split wood, it permits loading substantially more wood.
Ash overflows in ashbin.	You must empty ashbox periodically or it overflows. Clean it out with wrought-iron firesets, preferably one that includes a hoe.
Water in the stack.	Water vapor is a by-product of wood combustion, which condenses at roughly 212 degrees. When the smoke moves slowly up the chimney, or during a slow burn when the vapor's temperature out of the stove was already fairly low, the water vapor will lose heat and condense. Although dry wood still gives off moisture, it burns better. If problem continues, try to speed up the draft using a draft-inducing chimney cap, but be sure to have a flue pipe damper so that you will still have some control over the draft. The stove pipe and chimney should also be as straight as possible because turns in the smoke path accelerate the cooling process. Heating up the stack when you start a fire by running a hot fire for awhile, will help. Also, speed up the draft and dilute the creosote and water vapor by opening up the third air source—the one that directs air to the combustor.

Glossary

blank-off plate—A round plate that closes off a humidifier bypass during air conditioning operation.

blast tube—The round, pipe-shaped casting in which the oil nozzle and electrodes are located. Extends to the rear and supports the motor, blower, and oil pump. Also holds the air turbulator (swirler) to spin the air for the flame.

blower—A drum-shaped fan, as opposed to a propeller fan. The drum has louvered blades, spaced around the circumference and is placed in a sheet metal housing. Found on furnaces and air conditioners.

boiler trim—Necessary accessories attached to a boiler in addition to the casing, that make the boiler operational. Includes: gauge glass, try-cocks, low-water cutoff/feeder device, high limit, operating limit, and drain valve.

breaks—Disconnects electrical current, as turning Off a toggle switch.

B.T.U.—British Thermal Unit. The amount of heat required to heat one pound of water one degree (1°)Fahrenheit.

bushing, electrical—A smooth-surfaced fitting screwed onto a length of conduit to protect cable from abrasion at the end.

B-valve—A small, shutoff valve, generally used on conversion gas burners. The main shutoff valve, mounted on the supply line five feet off of the floor, has a tapped opening on the supply side of this valve. The B-valve is installed here so that the pilot gas supply can be turned on independent of the main gas supply. Thus the main gas valve can be shut off, then the pilot valve opened and the pilot can be lighted safely.

capillary tube—Small diameter tubing installed between the high and low sides of an air conditioning system. The action is similar to a thermostatic expansion valve.

cheater—An extension (pipe length), slipped over a pipe wrench handle to provide extra leverage.

circuit breaker (breaker)—An electrical device to protect wiring against overcurrents. Located in the electrical panel.

clean out, chimney—A small metal door at the bottom of a chimney for cleaning soot and debris from the chimney base. The condition of the chimney can be determined by holding a mirror at an angle inside the base.

cleat-drive—A sheet metal strip shaped so as to allow it to be driven onto turned-over edges of sheet metal ductwork to draw two sections together.

cleat, S-type—A sheet metal strip shaped into an S shape or Z shape to hold alternate sides of a duct together. Used with a drive cleat.

combustion chamber—A specially designed housing, in which a gas or oil flame is contained. May be constructed of stainless steel or hard-fired or insulating firebrick.

condensate—Moisture from ambient air that condenses on cold surfaces, such as an iced drink or other cold surface.

condensate pump—A small, fractional horsepower pump for lifting condensate from a cooling coil and pumping it to a remote drain. Used when condensate cannot drain by gravity to a nearby floor drain.

condensing unit—The part of an air conditioning unit located outside. Consists of the motor/compressor, condenser, condenser fan, cabinet, and related controls.

constant level valve—A float-type valve used to meter fuel oil to a pot-type radiant or circulating space heater. Can be manually or automatically controlled to regulate heat.

cooling coil, A-shape—A coil in the shape of an inverted ''V.'' Saves vertical height in plenum, as opposed to a flat coil.

dehumidifier—A device to remove moisture from an area, usually a basement, by a refrigeration system condensing moisture from the air.

differential—The difference between the On and Off settings of a control device. A fan control might have settings of, On at 120°F, Off at 90°F. In this case, the differential is 30°.

draft, negative—A draft over a fire in a fire pot that blows back out of the fire pot door when slightly opened. Can be checked by holding a lighted match near the slightly opened door.

draft, positive—A draft over a fire in a fire pot that sucks into the fire pot a lighted match flame held near a slightly opened door.

draft stabilizer—A hinged damper installed in a flue pipe tee directly after the flue connection of an oil burning appliance that provides a constant draft over the fire. The damper swings on hinges and has an adjustable weight, providing a ''stabilized'' draft over the fire.

electric resistance element—An electrical wire that heats up when an electrical current is passed through it such as a toaster or electric range surface, or oven element. Also used for electrical resistance space heating when gas or oil is not available. The element can be covered or coiled and stretched between porcelain insulators in ductwork.

electrodes—Rounded, metal end wires, supported by porcelain holders. Used in pairs to provide a spark between the two ends. The ends are separated 5/32 inch to provide the spark gap. The spark is used to light oil in an oil burner in a furnace or boiler.

end bell—Either end of a motor housing, held in place by long ''through'' bolts running completely through the motor housing.

end switch—A small switch on the motor of a zone hot-water valve that makes electrical contact when the valve opens, and opens contact when the valve closes. By making contact, the switch closes a relay which then turns on the circulating pump.

evaporator—The cooling coil located directly above the furnace casing in its own casing.

fan—Generally, a ''propeller'' type fan, not a blower wheel.

ferrule—A brass sleeve, part of a compression fitting for copper tubing. When the compression nut is tightened, the ferrule squeezes into the tubing to make a tight joint. Not recommended for either gas or oil lines.

firebrick, hard fired—A hard firebrick, similar to house face brick. Used in larger sized commercial installations.

firebrick, insulating—Similar to hard-fired brick, except much softer. Can be cut with a coarse hacksaw. Used for small residential installations.

fire valve—An automatic fuel line shutoff valve in an oil line to a burner. If the temperature rises in the vicinity of this valve, the valve snaps shut, stopping any oil spread.

flame-out—The disappearance of the flame in an oil burner combustion chamber from lack of oil or stoppage of the burner, due to malfunction.

flare—A cone-shaped widened end on a length of soft copper or aluminum tubing. Mates with the slanting surface of a flare fitting, a 45° angle.

flaring tool—A tool consisting of a clamping bar with a series of holes in its two-piece body. The holes have flared or chamfered tops. Clamps hold the two halves of the bar together after the tubing has been positioned in the proper hole. A flare mandrel hooks over the bar, and a cone on the mandrel flares the tubing end.

flexible connection—A connector made of flame-resistant fabric used between the furnace cabinet and the ductwork to prevent vibration.

flexible coupling—A coupling used on an oil burner, between the motor and fan, and pump. Usually made of rubber with separate or connected metal couplings. Three-piece coupling ends can be interchanged for others of different design or size.

float valve—A small valve controlled by a float, to maintain a constant level of water in a humidifier pan.

flue liner—A vitreous-fired clay lining, used to line brick chimneys to provide a gastight flue passage.

flue pipe—The sheet metal pipe connected from the heating equipment to the chimney.

fuel oil, No. 2 grade—Lightweight, nearly colorless oil from petroleum. Suitable for use with domestic pressure type oil burners.

furnace bonnet—The top section of a furnace, usually refers to a gravity furnace. Heat pipes for various rooms start here.

furnace, downflow—A forced, warm-air furnace, in which the air enters at the *top* and leaves from the *bottom*.

furnace, horizontal—A forced, warm-air furnace, in which the air flows horizontally. Used in crawl spaces or in an attic.

furnace, upflow—A forced, warm-air furnace, in which the air enters at the *bottom* and leaves from the *top* of the furnace.

fused disconnect, weatherproof—An electrical switch with fuses that is installed outdoors, close to the condensing unit.

handy box—An electrical junction box designed for holding a switch or outlet, or for making wire connections.

heat exchanger—The heavy sheet metal or casting that holds and contains the burner flame and separates the products of combustion from the circulating air from the rooms.

heat pump—A special heating and cooling system, in which the air conditioning system is arranged so that the indoor cooling coil and the outdoor condenser functions are reversed. The indoor cooling coil becomes the system condenser and is used for heating the house. The outdoor condenser becomes the cooling coil and absorbs heat, which is ''pumped'' into the indoor ''heating'' coil.

high limit control—A safety switch that operates to stop the burner (oil, gas, or electric) in the event of overheating, or other malfunction of the equipment.

humidifier—A device to add moisture to warm-air or hot-water heating systems. Consists of a rotating, foam-covered wheel that dips into a water reservoir, wetting the foam, then giving up the water to the air moving over the wheel.

I.D.—Internal diameter of pipes, hollow circular objects, etc.

ignition transformer—A transformer that steps up the voltage from 120V to 10,000V. Provides an electric spark to ignite the oil spray in an oil burner.

insulation—Materials, that when used in buildings, retard the gain or loss of heat. Special types are also manufactured to retard noise transmission.

intermittent pilot—A gas pilot that is ignited by an electric spark, on a call for heat from the thermostat. The pilot stays on the entire time the main burner is on.

internal switch—A small switch inside an electric motor used to disconnect the starting winding of the motor after the motor has reached normal speed on its running winding.

leak detector—A device to locate refrigerant leaks in a system. Various types are, soap solution, propane torch, and electronic.

lock out—The stopping of any mechanical or electrical device, by mechanical, heat, elapsed time, or other means. Then the equipment can only be restarted manually.

louver—A register having narrow strips or fins, usually in a pattern, used for ventilation, air delivery, or exhaust.

makes—Establishes electrical contact, as turning On a switch.

nipple—A short length of pipe, threaded on both ends.

O.D.—Outside diameter of pipes or other objects.

oil pump—The pump of an oil-burning furnace that draws oil from a tank and forces it through the oil nozzle at 100 PSI to produce an oil flame.

one pipe, oil burner—A piping arrangement, using only one oil line between the oil tank and the burner. The oil tank is usually slightly above the level of the oil burner.

pipe thread, female—Internal thread on a fitting.

pipe thread, male—External thread on a pipe.

pipe union—A pipe connection designed so that the pipe can be disconnected at the union location by unscrewing the union collar.

plenum—The section of duct directly leaving the furnace or air conditioner that smaller ducts branch off to various locations.

pre-charged refrigerant lines—Copper refrigerant lines, one large 7/8 inch O.D. suction line (insulated) and one small 3/8 inch O.D. liquid line. Each end has a special spring-loaded coupling that is connected to the two separate components of the system. When these couplings are connected to the mating couplings on the evaporator (inside), and on the condensing unit (outside), the internal connections are opened, resulting in a completely charged system.

prick punch—An extra ice pick type of pointed tool. Used for punching holes in light sheet metal or wood.

prove, a flame—To sense that a flame has been established in a combustion chamber. Methods of doing this are, by the increase of heat in the flue pipe (stack), or by an electric eye reacting to the flame.

PSIG—Pounds per square inch gauge pressure. Sometimes stated as PSI.

pulley, adjustable—A vee-type pulley, one side of which can be unscrewed from the other side. Screwing or unscrewing the moveable side increases or decreases the effective diameter.

reducing coupling—A pipe coupling with a smaller size thread at one end and a larger size thread at the other. Also can refer to elbows and tees.

relative humidity—The amount of water in the air as a percentage of the maximum amount that the air could hold at a given temperature.

relay, electrical—A device that closes electrical contacts when an electromagnetic coil is energized such as a 120V pump motor is started upon a signal from a 24V thermostat.

safety shutoff valve, 100%—A valve with a pilot connection designed in such a way that only the pilot can receive gas when it is being lighted. A knob must be held down during and after lighting the pilot, for about 30 seconds. Only then can the main gas valve be turned on. If the pilot goes out for any reason, the main gas and pilot are both shut off. This is the "100%" feature.

satisfied—When the surrounding temperature equals the thermostat setting and causes the thermostat to shut off the heating equipment.

seasoned, wood—Cut lengths of firewood that have been stored under cover for at least six months.

sensor—An electrical or bi-metallic part that moves or sends an electrical signal to operate some other device.

setback—Reducing the thermostat setting for night or when the house is unoccupied automatically by a clock thermostat that is so programmed.

set point—The thermostat setting, selected by the homeowner, such as 73°F., etc.

set-up—Raising the temperature of a house that is unoccupied to reduce the running time of the air conditioning. This is accomplished by a special thermostat.

short cycle—The repeated starting and stopping of a furnace or air conditioner much oftener than is normal.

shutdown—The stopping of operation of any mechanical or electrical device, by mechanical, heat, electrical, or other means.

shutoff valve—Any valve that is able to shut off the flow of a liquid or gas through a pipeline.

space heater, circulating—An oil or gas heater that has the hot surface of the fire pot covered by a casing, spaced away from the hot surface. This induces circulation into the space surrounding the heater.

space heater, radiant—An oil or gas heater that has the hot surfaces exposed to the touch and not insulated or shielded.

spark gap—The distance between the spark electrodes of an oil burner, 5/32 inch, across which a spark is formed. This ignites the oil sprayed into the combusion chamber.

stack relay—An electrical-mechanical device to control the operation of an oil burner in response to the stack temperature. A bi-metallic helix, a part of the control, is inserted into the stack and responds to temperature changes. If the stack heats up, the burner is allowed to continue to run. In this case, a flame has been established. If the stack does *not* heat up, a warp switch trips and ''locks out'' the burner. The relay must then be manually reset. This alerts the owner to find out the cause of the lock out.

standing pilot—A gas pilot that burns continuously, unless shut off manually.

subbase, thermostat—An extra base for a thermostat. Provides switching for various functions of a heating/cooling system, such as, *Heating - Off - Cooling* and *Fan - Auto - Off*.

supplementary heat—Extra heat needed to supplement the heat supplied by the heat pump when the system *cannot* maintain comfortable temperatures in the house. Can be oil, gas, or electric resistance heat.

Teflon tape—A white tape made of Teflon that is wrapped around male pipe threads instead of pipe thread compound.

thermostatic expansion valve—A device to meter the amount of refrigerant to properly fill the cooling coil (evaporator) so that full capacity of the system is obtained.

thinwall tubing—Electrical metallic tubing (EMT) is a thin walled light metal tubing used as electrical conduit that is joined with special connectors and couplings—it cannot be threaded.

try-cock—A small valve screwed into the side of a *steam* boiler that can be opened to determine the level of water in the boiler. Three try-cocks are installed in a vertical row at about four inch intervals, one above the water line, one below the water line and one at the water line. If they are opened one at a time it makes it easy to determine the true water line. Many times the gauge glass adjacent to the try-cocks does not give a true water level.

two pipe—A piping arrangement whereby two oil lines are run between the oil pump and the oil tank. Necessary for proper operation, if the oil tank is underground or at a great distance.

wire connectors—A small plastic or porcelain nutlike item used to connect two or more bared wire ends without soldering. Also called wire nuts.

zone valve—A special low-voltage valve for hot-water systems. The valve is controlled by a thermostat so that certain zones (sections) of a building can be temperature controlled independently of other zones. Most zone valves incorporate an "end switch" that turns on the circulating pump for the complete system when the valve opens. It should be noted that *any* valve is able to turn on the pump, but that *all* valves must be closed for the pump to stop.

Index

D

E

F

G

H

I